THE NEPTUNE FILE

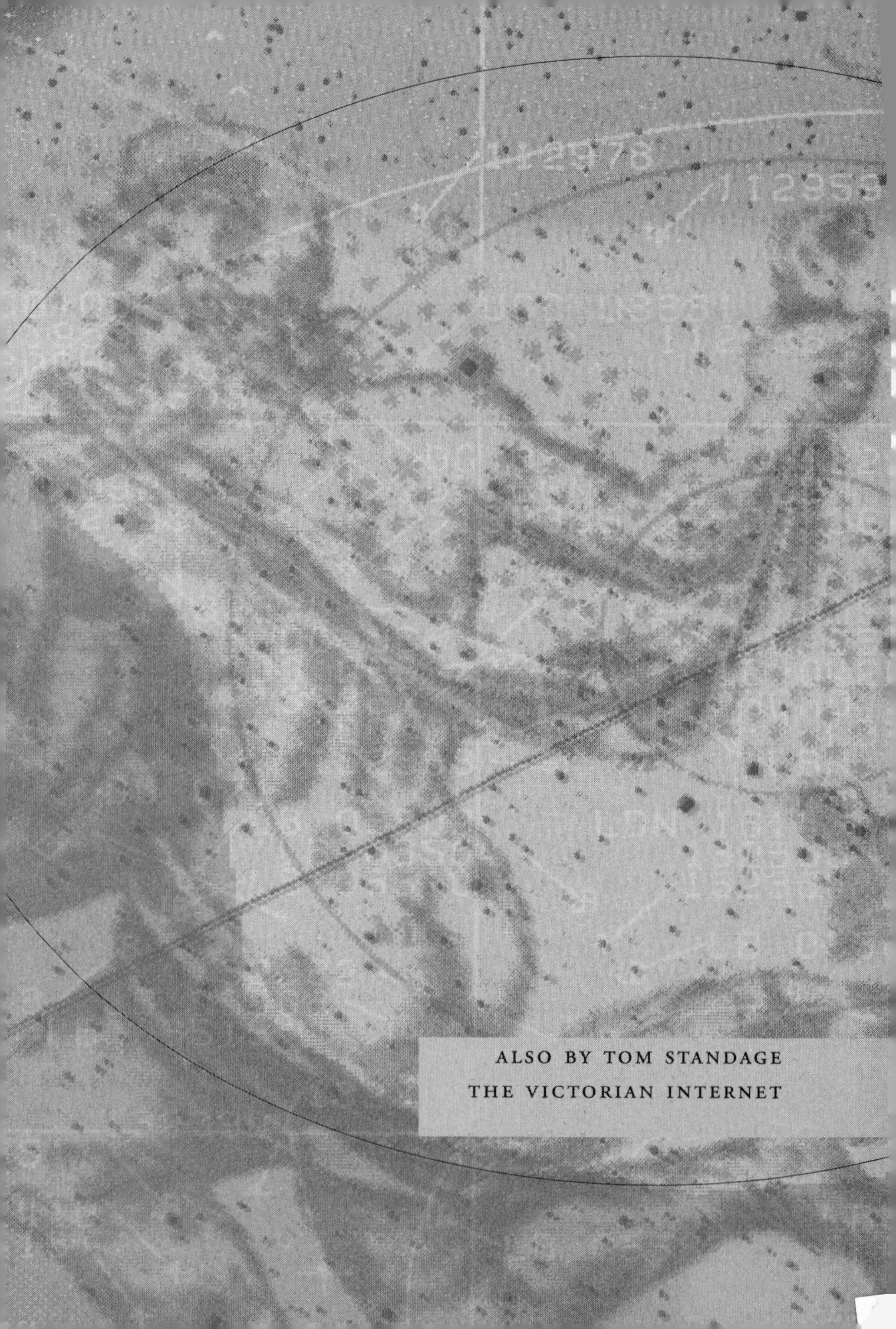

ALSO BY TOM STANDAGE

THE VICTORIAN INTERNET

THE NEPTUNE FILE

A STORY OF ASTRONOMICAL
RIVALRY AND THE PIONEERS
OF PLANET HUNTING

Tom Standage

WALKER & COMPANY

NEW YORK

TO MY PARENTS, WITH THANKS

First published in the United States of America in 2000 by
Walker Publishing Company, Inc.

Published simultaneously in Canada by Fitzhenry and Whiteside,
Markham, Ontario L3R 4T8

The author has made every effort to locate and contact all the
holders of copyrighted material reproduced in this book.

STAR MAPS IN THIS BOOK WERE PRODUCED USING STARRY NIGHT™
ASTRONOMY SOFTWARE. VISIT HTTP://WWW.SIENNASOFT.COM TO
LEARN MORE AND DOWNLOAD A FREE TRIAL VERSION. OR CONTACT:
SIENNA SOFTWARE, INC., 303-411 RICHMOND ST. E., TORONTO, ON-
TARIO, M5A 3S5, CANADA. TEL: (416) 410-0259, FAX: (416) 410-0359,
E-MAIL: CONTACT@SIENNASOFT.COM.

Library of Congress Cataloging-in-Publication Data
available upon request
ISBN: 0-8027-1363-7

Book design by M. Fadden Rosenthal/mspaceny

Printed in the United States of America

2 4 6 8 10 9 7 5 3 1

Contents

Preface

• On June 26, 1841, a young Englishman paid a fateful visit to a bookshop in the university city of Cambridge. John Couch Adams, a mathematics student with a particular gift for astronomical calculations, was an expert at determining the orbits of comets, the times of eclipses, and the motions of the planets. As he scanned the shelves of Johnson's, an academic bookshop just around the corner from his college, he came across a small, red volume containing a report written nine years earlier on the subject of "progress in astronomy." Intrigued, Adams took it down from the shelf and started reading.

The report's author was George Biddell Airy, one of England's leading astronomers, and the passage that caught Adams's eye was a reference to the planet Uranus. While the rest of astronomy was, according to Airy, proceeding smoothly—observations were becoming ever more precise, telescopes more powerful, and predictions more accurate—the planet Uranus was demonstrating a mysterious refusal to behave itself. "With respect to this planet a singular difficulty occurs," Airy declared.

The positions of the other planets—Mercury, Venus, Mars, Jupiter, and Saturn, all of which had been known to astronomers since ancient times—could be predicted with extraordinary precision. But the position of Uranus, which had been discovered only sixty-five years earlier and was the most distant known planet from the Sun, could not. Every attempt to calculate its orbit, and draw up a table of its future position in the sky based on that orbit, had failed. Each time, the same thing happened: Uranus would adhere to the predictions for a few years and would then slowly start to drift away from its expected course.

Admittedly, the deviations in question were tiny—only a couple of hundredths of a degree or so, equivalent to the diameter of a penny viewed from a distance of 100 yards—and could only be detected using specialized instruments. But to astronomers used to being able to predict the positions of other planets with astonishing precision, to within a thousandth of a degree, Uranus was the cause of profound embarrassment.

As he stood in the bookshop reading Airy's report, Adams was aware that the problem of Uranus had not gone away in the nine years since it was written. The young mathematician was suddenly seized with the urge to solve the mystery. A few days later he wrote himself a memorandum, declaring that he had "formed a design, at the beginning of this week, of investigating, as soon as possible, after taking my degree, the irregularities in the motion of Uranus, which are yet unaccounted for."

Adams had a hunch that the unusual behavior of Uranus was the result of the subtle gravitational tugging of an unseen, as-yet-undiscovered planet orbiting even farther away from the Sun. He planned to attempt something that had never been done before: to analyze records of the position of Uranus and, by comparing its predicted position with its actual position, pinpoint the location of the undiscovered planet using nothing more than mathematical analysis. It would then, he hoped, be a simple matter for astronomers to verify his results by turning a telescope toward the appropriate part of the sky and finding the new planet.

Astronomers were accustomed to using telescopes to observe heavenly bodies first, and mathematics to explain their motions later. The idea that a planet could be discovered through calculation, rather than observation, was extremely daring. In contrast to most mathematical problems, which concern esoteric theoretical constructions and whose solution is often of direct interest only to mathematicians, the solution of this particular puzzle would, Adams hoped, result in the discovery of an entirely new world.

But things did not turn out at all the way Adams expected. On the day he chose to embark on the search for a new planet, his fate became inextricably entwined with that of Airy. And when the planet (known today as Neptune) was eventually discovered, the two men found themselves at the center of one of the most acrimonious disputes in the history of astronomy—an international controversy that has continued to smolder ever since.

The story of Adams's search for an unseen planet, and the uproar that it caused, was widely known in the Victorian era, though it is almost forgotten outside astronomical circles today. It is, however, more than just an intriguing historical yarn. Adams's work signaled the beginning of a new era of planetary discovery—it marked the genesis of astronomers using mathematics to search for new planets by looking for their telltale gravitational influence on other bodies, rather than by observing them directly with telescopes.

In recent years this approach has led to an extraordinary series of discoveries. In 1995 two Swiss astronomers, Michel Mayor and Didier Queloz, detected a planet orbiting a Sun-like star in the constellation of Pegasus by analyzing tiny fluctuations in the star's light caused by the planet's gravitational influence. This was the first planet to be found orbiting another sun.

Since then, using the same technique, astronomers have found dozens more "extrasolar" planets orbiting other stars, and the search for new planets has become one of the hottest fields in modern science. Thanks to the work of Mayor, Queloz, American astronomers Geoffrey Marcy and Paul Butler, and many other astronomers and theorists, there are now charts of alien solar systems, theories to predict and explain the characteristics of these distant worlds, and plans to build ever more ambitious planet-hunting equipment. Every year the tally of new planets increases as the sensitivity of the instruments used to detect them is improved. Even though none of them has ever been seen, far more planets are now known to exist outside our solar system than within it.

In the light of these discoveries, the work of Adams and his contemporaries—George Airy, James Challis, François Arago, and Urbain Jean-Joseph Le Verrier—takes on a new significance. Their story has both inspired and haunted would-be discoverers of new planets ever since, and their pioneering work will form the basis of the next generation of planet-hunting techniques and technologies.

Here, then, is the story of Adams's attempt to detect a new planet using nothing more than mathematics, and of the modern-day and future planet hunters following in his footsteps. The influence of the elusive planet Neptune can be seen as a thread that runs through the entire history of planet hunting—an adventure that begins sixty years before Adams's visit to the bookshop, and is still unfolding today, over two centuries later.

Acknowledgments

• I AM GRATEFUL to many people for helping to make this book possible, particularly George Gibson and Jackie Johnson at Walker & Company, and Katinka Matson at Brockman Inc., for pointing me in the right direction. Peter Hingley at the Royal Astronomical Society, Adam Perkins at the Royal Greenwich Observatory archive in Cambridge, Françoise Launay at the Paris Observatory, Patrick Moore, Robert W. Smith, and Christopher A. Robinson provided invaluable guidance during my research. Brian Marsden at the Harvard-Smithsonian Center for Astrophysics, Michael Nieto, and Didier Queloz gave generously of their time and ideas. My thanks are also due to my colleagues Geoffrey Carr and Andreas Kluth at the *Economist* for putting up with my obsession with planetary astronomy and helping with impenetrable nineteenth-century German, respectively. I am grateful to Chester for his many helpful comments. In addition, I would like to thank Stefan McGrath, Judi Kloos, Virginia Benz and Joe Anderer, Oliver Morton, Anna Aebi, Philip Millo, the Marti family (Cristiana, Claudio, and Franca and all the cats, dogs, and toads of Montegualandro), Kathy at the Farmhouse Inn, and Oscar and Millie. Finally, I would like to thank my wife, Kirstin, for her patience and support.

THE NEPTUNE FILE

The Musician
of the Spheres

Then felt I like some watcher of the skies
When a new planet swims into his ken.

—JOHN KEATS
"ON FIRST LOOKING INTO CHAPMAN'S HOMER"

Sometime between ten and eleven o'clock on the night of Tuesday, March 13, 1781, William Herschel was looking at the stars through a homemade telescope from his garden in the English spa town of Bath. Herschel was a musician by trade, but his passion for astronomy had grown over the previous few years to the point that he was spending more time with his astronomical instruments than with his musical ones. What he saw through his telescope that night was to change his life completely and win him widespread and lasting fame. He was about to become the first person to discover a new world.

For centuries, astronomers had followed the five "classical" planets—Mercury, Venus, Mars, Jupiter, and Saturn—across the skies. The existence of these planets, which resemble bright stars clearly visible to the naked eye, has been known to humankind almost from the beginning of civilization itself. But since antiquity nobody had ever discovered any additional planets, and the idea that there might be more such objects lurking in the heavens seemed outlandish.

Herschel was in a unique position to make such a fortuitous discovery. Having taught himself astronomy, he had little interest in the tedious business of measuring the positions of the stars or working out tables of the positions of the Moon and planets, which was what professional astronomers spent most of their time doing. Instead, as an amateur, he was free to roam the skies at will, looking at whatever took his fancy. At the same time, Herschel was no ordinary amateur. As a result of the trial-and-error process of learning to build his own telescopes, he had, without realizing it, become the finest telescope maker in the world.

The telescope he was looking through on that chilly March night was one of his favorites: 7 feet long, 7 inches in diameter, with a wooden tube and a handmade mirror that was the result of hours of painstaking grinding and polishing. The tube was supported by an elaborate wooden frame, with a system of cords and pulleys and three small crank-handles to adjust its position. Herschel also had beside him a set of his own eyepieces, each one mounted in a tube of cocus wood, the kind of wood used in the body of an oboe—one of the first musical instruments he had learned to play as a boy. By removing one eyepiece and inserting another, Herschel could vary the magnifying power of his telescope.

Increasing the magnification would, for example, make a planet (such as Jupiter or Saturn) appear larger and more distinct. Herschel particularly enjoyed looking at Saturn, which, with its spectacular ring system, is a magnificent sight in even the smallest, feeblest telescope. But on that particular night he was looking at stars, not planets, using one of his less powerful eyepieces, with a magnifying power of 227 times.

As he swept the telescope over the constellation of Gemini, Herschel noticed something unusual. He decided to take a closer look and removed the telescope's eyepiece in order to substitute a more powerful one. Switching to the eyepiece with a magnification of 460, he found that his mystery object appeared twice as large as it had under a magnification of 227; when he switched to the eyepiece with a magnification of 932, it doubled in size again. Because stars are so distant that they appear as points of light, no matter how great the magnification, this meant the mystery object was definitely not a star. So he noted it in his astronomical journal as "a curious either nebulous star or perhaps a comet."

The object was essentially a slightly fuzzy blob. Herschel knew that a fuzzy blob could be one of two things: a nebula (a generic term covering all manner of star clouds, clusters, and what we now know are distant galaxies) or a comet (an orbiting snowball within our own solar system that brightens and spews out a tail of gas and dust as it approaches the Sun). The two can be told apart by seeing whether or not they move relative to the fixed stars. Nebulae, like stars, stay fixed; comets, like planets, move from one night to the next. Hoping that perhaps he had discovered a comet, Herschel noted its position so he could observe it after a few days and see if it was still in the same place. A few days later, on the night of Saturday March 17, he noted in his journal, "I looked for the comet or nebulous star, and found that it is a comet, for it has changed its place."

To have discovered a comet was quite an achievement, and Herschel knew what to do: He had to inform the astronomical community as quickly as possible, to establish his priority as the discoverer. In those days, the members of the worldwide scientific community informed each other of new discoveries, theories, and experiments via a constant blizzard of correspondence, often sending or receiving dozens of letters in a single day. So Herschel, who was only a peripheral member of this informal international network, immediately sent a letter containing the details of his comet to the most senior astronomer he knew: Thomas Hornsby, the director of the observatory in Oxford, with whom he had exchanged a few letters in the past. Through his friend William Watson, who moved in scientific circles in London, Herschel also informed Nevil Maskelyne, at the Royal Greenwich Observatory. Maskelyne was the astronomer royal, the most senior astronomer in the country.

Maskelyne found the comet almost immediately, and Hornsby found it a few days later. But they both realized that there was something highly unusual about it. "The last three nights I observed stars near the position pointed out by Mr Herschel, whereby I was enabled last night to discern a motion in one of them," Maskelyne wrote to Watson on April 4. But, he added, if this moving star was indeed a comet, it was "very different from any comet I ever read any description of or saw. This seems a comet of a new species." Maskelyne suggested that Herschel write a paper and send it to the Royal Society, the preeminent British scientific society, describing his telescope and his discovery.

Herschel's comet was unusual because unlike other comets, it had no tail and was not surrounded by a fuzzy cloudlike coma. In fact, it was hardly fuzzy at all. Maskelyne began to suspect that Herschel's comet was, in fact, an entirely new planet.

Most astronomers were not so sure. Herschel continued to observe what he assumed was a comet, and wrote up his results in a paper that Watson passed to the Royal Society in London, where it was read out loud at the April 26 meeting. Since London was several hours' journey from his home in Bath, Herschel did not attend. Modestly titled "Account of a Comet," his article nonetheless stirred up astonishment and skepticism in equal measure. For although Herschel's account of his discovery was straightforward enough, his casual reference to his eyepieces of 460 and 932 times magnification—along with two others of 1,536 and 2,010 times—astounded the astronomers present. Not even the astronomer royal's telescope, one of the finest available, could magnify any more than 270 times. So Herschel, with his wild claims about the power of his homemade telescope, sounded like a crank.

Crank or not, there was no denying that Herschel's comet was real and could be seen by any competent astronomer as it made its way through the heavens. Word of the comet soon reached astronomers overseas. Charles Messier, Maskelyne's counterpart in France and a senior figure at the Academy of Sciences (the French equivalent of the Royal Society), wrote to Herschel as soon as he heard of the discovery. A keen comet hunter himself, Messier was particularly impressed by Herschel's ability to spot such a small, faint object.

As summer approached and the evenings grew lighter, Herschel's comet was lost in the evening twilight and could not be observed again until August. By the time it reappeared in the darkening autumn skies, astronomers had started trying to calculate its orbit.

To begin with, they based their calculations on the assumption that the orbit was the usual shape for a cometary orbit, a mathematical curve called a parabola. Traveling on a parabolic orbit, a comet swoops in toward the Sun and then hurtles off again into the far reaches of the solar system. But working out a parabolic orbit for Herschel's comet that corresponded to its actual observed motion from night to night proved to be impossible. Even orbits that correctly predicted the comet's motion for a few days quickly became hopelessly inaccurate. Stranger still, the comet did not seem to be getting any larger or brighter, as comets normally do; indeed, Messier noted that with its small disk and whitish light, similar to that of the planet Jupiter, the comet was unlike any of the eighteen comets he had previously observed.

Anders Lexell, a celebrated mathematician and astronomer from St. Petersburg, Russia, decided to try a different approach. Instead of deriving a parabolic orbit, as would be expected for a comet, he derived the sort of orbit that would be expected of a planet. In 1609 the German astronomer Johannes Kepler had shown that the planets travel around the Sun in almost-circular ellipses. So Lexell performed a calculation to see if the motion of

Herschel's comet was consistent with a circular orbit. To his surprise, he found that it was. Furthermore, the orbit was far beyond that of Saturn, the most distant planet from the Sun. Lexell's results, and similar calculations performed soon afterward by other astronomers, tipped the balance of opinion in favor of the idea that Herschel had indeed discovered a planet— one whose faintness, due to its great distance, had prevented anyone from noticing it before.

This was a truly momentous discovery, and it prompted Sir Joseph Banks, president of the Royal Society, to write to Herschel in November 1781. "Some of our astronomers here incline to the opinion that it is a planet and not a comet," he declared. "If you are of that opinion, it should forthwith be provided with a name." If Herschel failed to move fast, Banks suggested, "our nimble neighbours, the French, will certainly save us the trouble of baptizing it."

In the same letter Banks also announced that the Council of the Royal Society had decided to award Herschel its highest honor, an annual prize called the Copley Medal, which Herschel was invited to London to receive. At the presentation ceremony on November 30, Sir Joseph made a speech praising Herschel for his discovery of a new planet and for having provided astronomers with a mysterious new body to observe, chart, and scrutinize. He then presented the medal, to great applause.

At the time he made his discovery Herschel was living a double life, combining music with astronomy. His journal entries contain an odd mixture of details of concerts, music lessons, and pupils one minute, and mirrors, glasses, putty, and star maps the next. He was obsessed; every spare moment was devoted to polishing mirrors, building telescopes, and observing the heavens. Often he would return from a concert or a social occasion in Bath and go straight to his telescopes. As his sister Caroline noted in her memoirs, "Every leisure moment was eagerly snatched at for resuming some work which was in progress, without taking time for changing dress, and many a lace ruffle was torn or bespattered by molten pitch."

Mirror making in particular is not a job for the halfhearted, since the mirror must be continuously polished for hours at a time in order to be free of imperfections. On one occasion, noted Caroline, "by way of keeping him alive I was even obliged to feed him by putting the Vitals by bits into his mouth. This was once the case when at the finishing of a seven-foot mirror. He had not left his hands from it for 16 hours altogether. And in general he was never unemployed at meals, but always at the same time contriving or making drawings of whatever came into his mind. And generally I was obliged to read to him when at some work which required no thinking."

One of Herschel's pupils, an actor named Bernard, recalled that one evening the sky began to clear in the middle of his music lesson. "There it is at last," cried a jubilant Herschel to the bewilderment of his pupil, dropping his violin and rushing to the telescope to observe a particular star. Bernard also described Herschel's rooms where the lessons took place: "His lodgings resembled an astronomer's much more than a musician's, being heaped up with globes, maps, telescopes, reflectors & c., under which his piano was hid, and the violincello, like a discarded favourite, skulked away in a corner."

Herschel himself recalled that some of his pupils "made me

William Herschel (*The Great Astronomers* by Robert S. Ball, London: Sir Isaac Pitman and Sons, 1895)

give them astronomical instead of music lessons." An invariably cheerful and good-tempered man, he was happy to oblige.

His unexpected discovery catapulted Herschel to international fame. Letters congratulating him were soon coming in from eminent astronomers all over Europe. The French astronomer Joseph-Jérome Lalande wrote from Paris to report that he and his colleagues at the Academy of Sciences, including Messier, had calculated their own approximate circular orbit for the planet and found it orbited the Sun roughly once every 80 years. Since he was writing a book on the history of astronomy, Lalande asked Herschel for information about both himself and his tele-

scope since, he said, astronomers would be "curious about everything which concerns you."

Many astronomers wanted to know more about Herschel's
extraordinarily powerful telescopes. "I congratulate you. . . . you
are the author of a truth which will make your name immortal
among Astronomers," wrote the Moravian astronomer Christian
Mayer, who went on to ask Herschel whether he would build
him a telescope and how much he would charge for it. Similarly,
Johann Schröter, a German astronomer based in Lilienthal,
wrote to ask if telescopes like the one with which Herschel made
his discovery were available for sale; if they were, he said, he
would like one, and so would his friend Johann Elert Bode in
Berlin. Yet another letter, from Georg Lichtenberg, a German
astronomer in Göttingen, praised Herschel as follows: "The accuracy of your observations is hitherto unheard of in astronomy.
It has given me special pleasure to see the courage with which
you undertake to examine afresh things which we had thought
finished and done with."

But some English astronomers still had their doubts about
Herschel. Was he genuinely an astronomer and telescope maker
of the highest order or merely a lucky amateur? In December
1781 Watson wrote to Herschel from London to point out just
how extravagant the claims he was making about his telescope
really were. By this time Herschel had sent a paper to the Royal
Society on the subject of double stars, in which he happened to
mention the magnification of his most powerful eyepiece, which
was over 6,000. This was simply too much for some of the astronomers present. "What! say your opposers," Watson explained, "opticians think it no small matter if they sell a telescope
which will magnify 60 or 100 times, and here comes one who

pretends to have made some which will magnify above 6000 times! Is this credible?"

To make matters worse, Herschel had also claimed that many stars that appeared as single stars to other astronomers were revealed by his telescope to be double stars. But nobody else could verify his claims. Was this because their telescopes were inferior, or did Herschel's suffer from some kind of optical flaw? Perhaps, Watson suggested, Herschel should invite other astronomers to examine his telescopes in order to confirm or disprove his claims about their quality. There was even muttering in some quarters that Herschel's planet might yet turn out to be a comet after all.

Any lingering doubts that Herschel really had discovered a planet were, however, finally demolished in the spring of 1782. By this time a number of astronomers had started working out more accurate orbits for Herschel's planet. The key breakthrough came when Bode discovered that a star, observed by the astronomer Tobias Mayer in 1756 and recorded in his star catalog, had subsequently vanished. The new, more accurate orbits could be used to see where in the sky Herschel's planet would have been at the time. The result was conclusive: Mayer's missing star was exactly where the planet was predicted to have been. Mayer had, without realizing it, seen the planet and mistaken it for a star. Herschel's discovery was thus confirmed beyond all doubt.

As the importance of Herschel's discovery became generally accepted, Sir Joseph Banks set about arranging a meeting between Herschel and the king. George III could be expected to look kindly upon Herschel; the king was known to be very interested in astronomy, and he and Herschel were both of Hanoverian origin and had been born in the same year. The right word in the right ear, Banks knew, could provide Herschel with the financial support to give up music and become a full-time astronomer, and thus make the most of his obvious astronomical talent.

Herschel soon heard from several of his friends in London that the king was interested in meeting him. On May 10, 1782, he received a letter from Colonel Walsh, a friend who knew George III directly, informing him that the king had asked when Herschel would be in London. Herschel decided to act. He packed his favorite 7-foot telescope into a carriage, together with his star atlases and catalogs, and set out for London with two aims in mind: to meet the king and convince the skeptics of the power of his telescopes.

Later that month Herschel met the king at Greenwich and presented him with a sketch of the solar system—its outermost extremity marked by the orbit of his new planet. The two men got on well, and the king arranged with Herschel that they should meet again at Richmond palace a few weeks later. In the intervening period, Herschel stayed in London and attended court events in Greenwich. "I pass my time between Greenwich and London agreeably enough, but am rather at a loss for work that I like," he wrote to Caroline. "Company is not always pleasing, and I would much rather be polishing a speculum."

Having made a good impression with the king, Herschel turned to his other business. He took his telescope to the Royal Greenwich Observatory so that it could be scrutinized by Maskelyne, the most senior astronomer in the country. If he could convince Maskelyne of the quality of his telescope, Herschel was sure that other astronomers would put aside their doubts.

The telescope was set up alongside the finest telescope in the Royal Observatory for comparison. Peering through the eyepiece, the astronomer royal was stunned to find that everything Herschel had claimed about his telescope was true, as Herschel explained in a jubilant letter to Caroline. "These last two nights I have been star-gazing at Greenwich with Dr Maskelyne," Herschel wrote to his sister. "We have compared our telescopes together, and mine was found very superior to any of the Royal Observatory." He also explained that, upon seeing the unusual stand Herschel had devised for his telescopes, Maskelyne had decided to order a copy of it to be made for his own use. But when he realized that the Royal Observatory's best telescope was hopelessly inferior to Herschel's homemade one, he changed his mind. "He is," wrote Herschel, "now so much out of love with his instrument that he begins to doubt whether it *deserves* a new stand."

Herschel also visited Alexander Aubert, an amateur astronomer who owned several especially fine telescopes. In particular, Aubert had one made by James Short, who was generally acknowledged to be the finest telescope maker in the country. In a letter to his brother, Herschel reported that Aubert's telescopes "would not at all perform what I had expected, so that I have no doubt that mine is better than any Mr Aubert has; and if that is the case I can now say that I absolutely have the best telescopes that were ever made."

Herschel had been triumphantly vindicated. Although he was self-taught, he had become not only an accomplished observer but the maker of the finest telescopes of his day. Even more impressively, he had become the first astronomer to discover a planet, one several times larger than Earth—larger, indeed, than

any of the other planets except Jupiter and Saturn. Now it
needed a name.

When it came to bestowing a name upon Herschel's planet,
there was no shortage of suggestions. In France, Lalande proposed that the planet be named after its discoverer, following
the practice of botanists and anatomists. But other astronomers
thought the name Herschel would look out of place, given that
all the other planets were known by mythological names. Which
mythological name would be most appropriate for the new
planet?

Lichtenberg, who was well known in Germany as a satirist,
sarcastically proposed the name Astrea, after the goddess of justice. Since she had so manifestly failed to establish her reign on
Earth, he reasoned, perhaps she had fled in anger to the most
remote corner of the solar system. Another suggestion came from
Louis Poinsinet de Sivry, a French scholar who proposed the
name Cybele, after the wife of Saturn. He argued that since
the fathers of the gods, Jupiter and Saturn, were represented in
the heavens, why not install the mother too? The list went on.
Other suggestions included Hypercronius (literally, "above Saturn"), Minerva (after the Roman goddess of wisdom), and,
given that the new planet's orbit defined the boundary of the
solar system, Oceanus (after the mythical river thought to surround the Earth). But none of these suggestions got very far.

Erik Prosperin, astronomer to the king of Sweden, was taken
slightly more seriously when he suggested the name Neptune.
This would, he said, make sense from a mythological point of
view: According to Roman mythology Neptune, the god of the

sea, was also the brother of Jupiter; both were the sons of Saturn, who would then be found in the heavens orbiting the Sun between his two sons. Following Prosperin's lead, and as one of the first to have realized that Herschel's discovery was truly a planet, rather than merely a comet, Anders Lexell felt entitled to contribute his own suggestion. He suggested the name Neptune of George III or Neptune of Great Britain, as a tribute to British naval supremacy, following the defeat in April 1782 of the French fleet by Admiral Rodney off Dominica. But other astronomers thought this a strange suggestion; the planet had, after all, been discovered by a Hanoverian, and its planetary nature had been confirmed by the work of continental, not British, mathematicians. In any case, Sir Joseph Banks had been proved right: Because William Herschel had failed to propose his own name for the new planet, it was now open season.

On July 2, 1782, Herschel went to visit the royal family at Windsor, and the following day he wrote to Caroline: "Last night the King, Queen, the Prince of Wales, the Princess Royal, Princess Sophia, Princess Augusta, &c., saw my telescope and it was a very fine evening. My instrument gave a general satisfaction; the King has very good eyes and enjoys observations with the telescopes exceedingly." Banks and his friends continued to work behind the scenes to encourage the king to provide Herschel with some kind of official recognition, and later that month their efforts paid off: Herschel was appointed private astronomer to the king, with an annual pension of £200, enabling him to give up his musical obligations and devote himself to astronomy. His new post carried no official duties, other than paying occasional visits to Windsor to make observations with the king, and although £200 a year was less than Herschel was earning as a musician, it was enough for him to live on. (At the time, even the astronomer royal's annual salary was only £300.) The king also paid Herschel to make him several telescopes so that he could present them as gifts. The Herschels subsequently moved from Bath to Slough in order to be nearer the king at Windsor.

The king's action was widely applauded, and Herschel's friends and patrons were thrilled at his success. In France, despite the animosity between the two countries, Lalande praised the king's enthusiasm for spending money on telescopes rather than on waging war. (There was some truth in this; but after thumping the French navy, Admiral Rodney had in fact received a pension worth ten times as much as Herschel's.) Inevitably, the king's generosity influenced Herschel's opinion about what to call the new planet. Soon afterward Herschel wrote an official letter to Banks proposing that it should be known by the Latin name Georgium Sidus, which, somewhat confusingly, means "the Georgian star."

This name, Herschel explained, was appropriate for a number of reasons. First, he believed that the mythological names favored by the ancients were inappropriate in the modern scientific age. Instead, he declared, "the first consideration in any particular event or remarkable incident, seems to be its chronology; if in any future age it should be asked, when this last Planet was discovered? It would be very a satisfactory answer to say, 'In the Reign of King George the Third'. As a philosopher then, the name Georgium Sidus presents itself to me, as an appellation which will conveniently convey the information of the time and country where and when it was brought to view."

As he paid tribute to his patron, Herschel became rather carried away. "As a subject of the best of Kings, who is the liberal protector of every art and science; as a native of the country from whence this Illustrious Family was called to the British throne; and as a person now more immediately under the protection of this excellent Monarch and owing everything to his unlimited bounty; I cannot but wish to take this opportunity of expressing my sense of gratitude, by giving the name Georgium Sidus to a star which (with respect to us) first began to shine under His auspicious reign."

In addition to acknowledging the king's generosity, dedicating the planet to him had a political overtone, as the scientist and lecturer Matthew Turner was swift to point out. In October 1781, just as Lexell's calculations confirmed the planetary nature of Herschel's discovery, the British forces in America were surrendering at Yorktown. "It is true we had lost the *terra firma* of the Thirteen Colonies in America," Turner noted, "but we ought to be satisfied with having gained in return by the generalship of Mr Herschel a *terra incognita* of much greater extent *in nubibus*."

Herschel's suggestion was certainly politically expedient, but it was greeted with groans from continental astronomers. Lexell objected that Georgium Sidus, as well as being unwieldy, was inaccurate because Herschel had discovered a planet, not a star; Lalande continued to insist that Herschel was still the most appropriate name. Had Herschel suggested a more practical name, his fellow astronomers no doubt would have fallen into line, but it turned out to be Bode, astronomer to the Royal Prussian Academy of Science in Berlin, who made the decisive move.

"I have the honour to congratulate you heartily on the fortunate discovery of the new planet," Bode wrote to Herschel. He went on to say how proud he was to have been the first person in Germany to have observed it, on August 1, 1781, "since when I have observed it as often as opportunity offered, and have published my observations in my Astronomical Yearbook." This was a set of astronomical tables that Bode published every year, which meant it was effectively up to him to decide how the new planet would be referred to within Germany. And fortunately Bode had decided on a sensible name. "I have," he wrote, "proposed the name Uranus for the new planet."

Bode's suggestion made more sense than any of the others. Uranus was the father of Saturn and grandfather of Jupiter, and the name also had astronomical overtones, since Urania, one of the nine Greek Muses, was the patron of astronomy. Bode explained that he had chosen the name Uranus as he thought it better to stick to mythology. But, he explained to Herschel, "I do assure you that had I been in your situation I should have felt it proper to do as you have done." Abbé Maximilian Hell, the director of the Vienna Observatory, wrote a poem in Latin commending Bode's suggestion and then used the name in the planetary tables published by his observatory, so that it was swiftly adopted in Austria too.

And so Herschel's new planet ended up with three names. It was known in Britain as the Georgian planet or simply the Georgian; in France as Herschel; and in Germany, Austria, and elsewhere as Uranus, the name by which it is universally known today.

Whatever its name, the new planet was the cause of great excitement among astronomers. In his speech on presenting Herschel with the Copley Medal, Sir Joseph Banks had speculated about the discoveries that would result from Herschel's new planet. "Who can say what new rings, new satellites, or what other nameless and numberless phenomena remain, waiting to reward future industry?" he wondered. His words were to prove oddly prescient.

2

Something Rather Better than a Comet

Mathematicians are like lovers. Grant a mathematician the least principle, and he will draw from it a consequence which you must also grant him, and from this consequence another.

— Bernard le Bovier de Fontenelle

A planet is, by definition, an unruly object. Unlike stars, which appear to be fixed in the sky in the rigid patterns of the constellations, planets move around. The five planets known to the ancients—Mercury, Venus, Mars, Jupiter, and Saturn—each resemble a bright, stationary star on a particular night, but by comparing the positions of the planets relative to the constellations from one night to the next, ancient astronomers soon realized that the planets' movements are highly complex.

Broadly speaking, planets move across the sky in a west-to-east direction, always keeping within a belt of constellations, the zodiac, which is strung around the sky. But rather than processing around the zodiac in an orderly fashion, the planets speed up and slow down. They may stop moving against the background of stars for days at a time or retrace their steps, moving backward from east to west. They may also move up and down slightly within the zodiac, thus tracing elaborate, elongated loops in the skies. The Greeks decided that the best word to describe these five capricious and unpredictable stars was *planetes*, which means "wanderer."

William Herschel's discovery of a new planet electrified the astronomical community. Not only was it an astonishing achievement in its own right; it also overturned the idea that all planets had been found, and raised the intriguing possibility that further planets might be lurking in the skies, waiting to be discovered. As Herschel's example showed, whoever found one would soon achieve international renown.

Why had Uranus not been identified as a planet before? It is, in fact, just visible with the naked eye, provided you know exactly where to look. But since it is so distant, its movement from night to night against the starry background is almost imperceptible, so it had escaped the notice of sky watchers for thousands of years. Even the advent of the telescope had failed to bring the dim, distant planet to the attention of astronomers—because in all but the most powerful telescopes, it looks exactly like a faint star. Indeed, that was exactly what Tobias Mayer thought it was when he observed the planet in 1756. This suggested that other undiscovered planets might already have been unwittingly observed by astronomers and mistaken for stars.

The Russian astronomer Anders Lexell declared that he now suspected the existence of several more planets even farther from the Sun. And the Moravian astronomer Christian Mayer confided to Herschel that he was certain there were a large number of planets among what had been assumed to be fixed stars. Attention turned to the question of where the next planet would be found, and some astronomers believed that Uranus could help point the way. One of them was Baron Franz Xaver von Zach, a German who paid William Herschel a visit in 1784 in order to observe the great astronomer at work.

Herschel himself was more interested in stars than in searching for additional planets. Indeed, he had discovered Uranus while surveying heavens in order to draw up a list of double stars. He planned to use repeated observations of double stars as a means of measuring their distance from the Earth, using a geometric method that had originally been proposed in the seventeenth century by the Italian astronomer Galileo Galilei. Herschel was also interested in investigating the fuzzy patches known as nebulae; he suspected that, through a sufficiently powerful telescope, they would be revealed as very distant clusters of individual stars. From these kinds of observations, Herschel hoped to advance astronomers' understanding of the structure of the universe. With this aim in mind, in 1785 Herschel hatched the plan to build the world's largest telescope in his back garden. This monstrous contraption would be 40 feet long and 4 feet in diameter, and its unprecedented light-gathering power would enable Herschel to view fainter celestial objects than had ever been seen before.

The king, who liked the sound of this plan, gave Herschel £2,000 to begin construction and subsequently provided another £2,000, plus £200 a year for the telescope's upkeep. The king also granted Caroline Herschel an annual pension of £50 for her work as her brother's assistant. Caroline Herschel was, in fact, starting to make a name for herself as an astronomer in her own right. In August 1786 she had made her first significant discovery—of a new comet—and she went on to become well known as a successful comet hunter.

Caroline Herschel (*The Great Astronomers* by Robert S. Ball, London: Sir Isaac Pitman and Sons, 1895)

The huge new telescope was soon taking shape in William's workshop. The mirror was 4 feet across and weighed nearly a ton, and was ground into shape by two shifts of twelve men in numbered uniforms on what one eyewitness described as "a sort of altar." This work was supervised around the clock by Herschel himself. In addition to the mirror, the telescope required the construction of a huge tube. At one point the king and queen, accompanied by the archbishop of Canterbury, paid a visit to see how construction was going, and Herschel invited his visitors to walk through the tube as it lay on the ground. The archbishop hesitated, prompting the king to extend his hand and say, "Come, my Lord Bishop, I will show you the way to heaven."

The telescope was finally completed on August 28, 1789. On its first night of use, Herschel turned the telescope on his favorite planet, Saturn, and discovered a previously unknown sixth moon.

The record-breaking telescope became quite a tourist attraction and was even referred to as the eighth wonder of the world. Actually, it was something of a white elephant; its large size made it rather unwieldy, and Herschel found it less useful than expected. He later commented that he believed 25 feet was the ideal length for a large telescope, because the large mirror in the 40-foot model tended to get fogged with dew or covered in ice on cold nights. Telescope making had become a valuable source of income for Herschel; he built the king of Spain a 25-foot telescope for £3,150, for example, and his records also show that he sold two smaller telescopes to the prince of Canino for £2,310—an absolute fortune at the time.

As well as receiving visits from the king and his guests and assorted members of the public, Herschel played host to most of

William Herschel's 40-foot telescope. (The Science Museum/Science and Society Picture Library, London)

Europe's greatest astronomers at one time or another. One guest recorded that "visitors often take unwarranted advantage of his courtesy and compliance, wasting his time and putting unnecessary and often ridiculous questions, but his patience is inexhaustible and he takes these inconveniences in such good part that no one could guess how much they cost him."

Despite the constant stream of visitors, Herschel maintained

his punishing observational schedule and continued to make discoveries. In January 1787, for example, he discovered that Uranus
had two moons. He never gave them names, but they are known
today as Oberon and Titania.

In 1788, Herschel married Mary Pitt, the widow of his friend
and neighbor John Pitt, a wealthy London merchant. William
proposed that he and Mary live at her house, with Caroline remaining at the Herschels' residence in Slough. But Mary realized
that William would not want to spend much time away from his
telescopes and workshop in Slough and insisted that she move in
with him, rather than the other way around. In 1792, Mary gave
birth to the couple's only child, a son they named John.

During his stay with the Herschels, Baron von Zach was
struck by the efficiency of the way in which William and Caroline Herschel made observations together. William would stand
outdoors at the telescope, while Caroline sat inside at her desk
by an open window, with a lamp, a ticking clock, and a copy of
John Flamsteed's star atlas in front of her. As her brother called
out his observations, she would write them down. Von Zach also
noted Herschel's strong constitution and the unusual way in
which he protected himself from catching a cold in the dampness
before dawn: "When the waters were out round his garden, he
used to rub himself all over, face and hands, etc, with a raw
onion, to keep off the infection of the ague."

The baron had his own reasons for wanting to pick up observing tips from the great Herschel. Von Zach was obsessed with
the idea of discovering a new planet for himself, and he thought
he knew where to find it. The evidence for its existence came
from an old mathematical rule of thumb that had, since the discovery of Uranus, taken on a dramatic new significance.

Von Zach was one of many astronomers who believed that there was a hidden pattern in the spacing of the planetary orbits. The origins of the pattern dated back to 1723, when Christian Freiherr von Wolf, a German philosopher, determined that if the average radius of the Earth's orbit around the Sun is taken to be 10 units, then the average radius of Mercury's orbit is 4, of Venus 7, of Mars 15, of Jupiter 52, and of Saturn 95. (Each unit is equal to about 9.3 million miles, and the average radius is used because planetary orbits are elliptical, so each planet's distance from the Sun varies slightly as it moves around its orbit. The distance of the Earth from the Sun, for example, actually varies between 9.8 and 10.2 units.)

In 1766 a Prussian scientist, Johann Daniel Titius von Wittenberg, proposed an explanation for these planetary spacings, which he slipped as a footnote into a book he was translating. Titius had noticed that the numbers conformed to a simple mathematical rule: Starting with 4 units for the radius of Mercury's orbit, it was necessary to add 3 for the radius of Venus's orbit, twice as much (2×3) for the radius of Earth's orbit, four times as much (4×3) for the radius of Mars's orbit, and so on, doubling the amount added each time. (See, page 27.)

There was just one problem: the huge gap between Mars and Jupiter. According to Titius's rule, the planet after Mars should have had an orbit with a radius of 28 units. But the planet in question, Jupiter, actually has an orbit with a radius of 52 units— the exact value predicted by the next term in the series. At 28 units, instead of a planet, there was a gap.

Admittedly, Titius was not the first to speculate that this gap might indicate the existence of a body unknown to astronomers. As early as 1761 Johann Heinrich Lambert, a Swiss-German phi-

Planet	Radius of Orbit	Predicted Radius
Mercury	4	4 + 0 = 4
Venus	7	4 + (1 x 3) = 7
Earth	10	4 + (2 x 3) = 10
Mars	15	4 + (4 x 3) = 16
?		4 + (8 x 3) = 28
Jupiter	52	4 + (16 x 3) = 52
Saturn	95	4 + (32 x 3) = 100

losopher, had commented, "Who knows if there are not planets missing in the large distance between Mars and Jupiter that will be discovered?" But Titius's theory went farther; his formula predicted that the missing body would be orbiting the Sun at a distance of 28 units, "where neither a chief nor a neighbouring planet is to be seen."

Titius was sure there had to be an underlying pattern to the structure of the solar system, and that he had uncovered it. He did not suggest the existence of an undiscovered planet in the space between Mars and Jupiter. After all, at the time, fifteen years before Herschel's discovery of Uranus, no new planet had been discovered in the history of astronomy. So instead Titius speculated that the space might be occupied by what he called a "neighbouring planet," or a moon orbiting Mars or Jupiter, despite the fact that such a moon would never have been able to stray so far into the gap.

Johann Elert Bode (the German astronomer who would later propose the name Uranus) came across Titius's suggestion in 1772 as he was completing a book of his own, and decided to

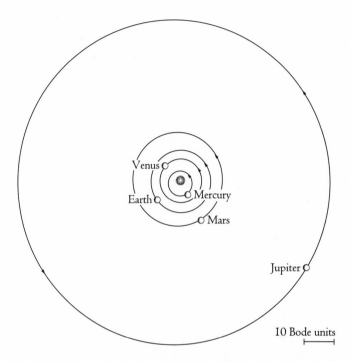

The orbits of Mercury, Venus, Mars, Earth, and Jupiter. The orbits of the outermost planets, Saturn and Uranus, are too large to appear at this scale. Planets are not shown to scale. Note the large gap between the orbits of Mars and Jupiter.

include it. He lifted Titius's text almost word for word, without acknowledgment, so that the theory came to be known among astronomers as Bode's law. Bode's only change was to omit Titius's spurious suggestion that the space might be filled by a moon; Bode simply wrote "planet." In a later edition of his book Bode mentioned Titius and also calculated that the missing planet ought to take 4.5 years to orbit the Sun.

Bode's law was, however, regarded as little more than a mathematical curiosity until the discovery of Uranus. According to the law, the orbital radius of the next planet beyond Saturn ought to have been 196 units. And, lo and behold, the average radius of Uranus's orbit was 192 units—near enough, given that the precise value was still uncertain. Suddenly, Bode's law looked as if it might be more than just a coincidence. In that case, there might really be a planet waiting to be discovered between Mars and Jupiter.

Yet Bode's law was no help in deciding exactly where in the skies to look for the hitherto unknown planet. The other known planets, including Uranus, all moved through the constellations of the zodiac, however, so that was the logical place to start. Furthermore, any planet between Mars and Jupiter ought to be close enough to Earth that it would be easily visible through a telescope. In 1787, having landed the job of court astronomer to Ernest II of Saxe-Coburg-Gotha, von Zach began a methodical search along the zodiac for the planet by looking for stars whose positions changed from night to night. He searched for thirteen years but found nothing.

So in the autumn of 1800 von Zach decided to adopt a more rigorous approach. At a meeting with five other like-minded astronomers at Lilienthal in Hanover, a planet-hunting club was founded. At the same time, von Zach launched a monthly astronomical journal, which would enable the members of the club to keep in touch. In his account of the club's founding, published in this journal, von Zach jokingly referred to its members as *Himmels-Polizey,* which means "celestial police."

The idea was that twenty-four astronomers from across Europe would look for the missing planet between Mars and Jupiter by dividing the band of the zodiac into twenty-four zones. Each astronomer would be allocated his own zone to patrol, much as a police officer walks a particular beat. This would hugely increase the chances of finding the elusive missing planet, if it existed, since each astronomer could get to know the positions of the stars in his zone and would then be far more likely to notice the appearance of an unknown planet. But things did not quite go according to von Zach's plan. Just as he was about to send out the letters to the twenty-four chosen astronomers to inform them of this scheme, one of them—Giuseppe Piazzi—made an unexpected discovery.

At the time Piazzi, who was based in Palermo, Sicily, was several years into the process of drawing up a new catalog giving the exact positions of over 6,000 stars. (Like von Zach, Piazzi had also visited Herschel in England, though his trip was spoiled somewhat when he fell off a high wooden ladder while looking through one of Herschel's mammoth telescopes and broke his arm.) Compiling the catalog involved observing the same stars, in batches of fifty at a time, on four consecutive nights, so that he could be sure of their exact positions. This approach also meant that the movement of any of the stars from one night to the next would be instantly apparent.

On January 1, 1801, Piazzi recorded the position of a faint star in the constellation of Taurus in one of his batches of fifty. But the next night, when he came to check its position, he found that it had moved. The possibility that he had made an error was ruled out on the third night, when he found that it had moved again. Piazzi surmised that he had discovered a new planet.

Like Herschel before him, Piazzi did not use the word "planet" immediately. On January 24 he wrote to Bode, Lalande, and his friend Barnaba Oriani, an Italian astronomer, telling them of his discovery, which he referred to as a comet. Only in the letter to Oriani did Piazzi admit that he actually thought it was "something rather better than a comet"—in other words, a planet.

Piazzi continued to track the new body until February 11 but was prevented by illness from making further observations. By the time his letters reached the other astronomers, the new body had moved too close to the Sun to be seen, so none of them could confirm his discovery. But that did not stop von Zach from jumping to the conclusion that his missing planet had at last been found. He immediately published an article in his monthly astronomical journal, titled "On a Long Supposed, Now Probably Discovered, New Primary Planet of Our Solar System Between Mars and Jupiter." Von Zach used the term "primary planet" to emphasize that the new body had been found orbiting the Sun in its own right; it was not merely a moon in orbit around one of the known planets.

Once Piazzi had made details of his observations available, the German astronomer Johann Karl Burckhardt, a former pupil of von Zach's, was one of several astronomers to work out an orbit for the new body. Von Zach published the orbit in July under the heading "Continued Reports About a New Primary Planet," together with a table of the planet's expected positions, and in September he published Piazzi's observations in full. But in the October edition he reported that, despite having looked for Piazzi's planet since its supposed emergence from the Sun's glare in August, astronomers had been unable to find it.

The problem was that Piazzi was the only person to have seen the new body, and he had made only a few days' worth of observations, so calculating an entire orbit was highly error-prone since it had hardly moved any distance in the sky during that time. Uranus, on the other hand, had been tracked by many astronomers for several months after its discovery, and Tobias Mayer's 1756 observation had been identified soon afterward, so it had never been in any danger of getting lost. But there were no such older observations for Piazzi's discovery, so an accurate orbit was much harder to calculate. Nonetheless, desperate to ensure his beloved planet's rediscovery, von Zach published a table of the new body's predicted positions for the months of November and December in his journal, so that his readers could assist in the search. Fortunately, help was to come from an unexpected quarter: the brilliant mathematician Carl Friedrich Gauss.

Gauss, who is sometimes called the "prince of mathematics," had an extraordinary talent that was recognized at an early age. When he was seven his schoolmaster told the class to add up the numbers from 1 to 100, thinking that it would keep the students quiet for a while; but Gauss immediately wrote the answer—5,050—on his slate. (He had noticed that the sum consisted of fifty pairs of numbers, 1 + 100, 2 + 99, 3 + 98, and so on, and that each pair added up to 101, so the answer was 50 × 101.) Gauss subsequently went on to study mathematics at the University of Göttingen and became particularly interested in the mathematics of astronomy. He worked on lunar theory and independently discovered Bode's law for himself. And in 1801, at the age of twenty-three, he came across the September edition of von Zach's journal, read about the lost planet, and decided to try his hand at finding it.

Rather than picking up a telescope, however, Gauss picked up his pen and spent the next two months working out some entirely new methods for calculating orbits. He had, in fact, been contemplating exactly this subject before he heard about the lost planet. Here was the ideal opportunity to try out his ideas.

Using Piazzi's observations and his new mathematical methods, Gauss swiftly calculated where the missing object ought to be. He sent his results to von Zach, who published them, along with similar calculations by other theorists. Finally, on the night of New Year's Eve, von Zach relocated Piazzi's discovery in Virgo—almost exactly where Gauss had said it would be. The following night it had moved. There was no doubt about it; the lost planet had been found.

Gauss's calculation of the orbit had proved to be the most accurate. To von Zach's delight, it put the average radius of the new planet's orbit at 27.67 units, which was very close to the 28 units predicted by Bode's law.

The rediscovery of Piazzi's lost planet highlighted Gauss's mathematical prowess and made his name as a mathematician. His newly derived methods had, he claimed, enabled him to calculate in one hour an orbit that would previously have taken three days. Indeed, in 1735 the great mathematician Leonhard Euler had worked continuously for three days to work out the orbit of a comet in order to win a prize competition, and he had gone blind in one eye as a result of the strain. "To be sure," declared Gauss, "I would probably have become blind also, if I had been willing to keep on calculating in this manner for three days."

Gauss was such a perfectionist, though, that it was not until 1809 that he was sufficiently satisfied with his refinement of orbital techniques to publish them in a book, *Theoria Motus*. Traditionally, mathematicians would guess at a rough orbit and then tinker with it by trial and error to minimize the errors between prediction and observation. Such tinkering was necessary because even the most careful astronomical observations are inevitably slightly inaccurate and are therefore mathematically inconsistent with each other. In his book, however, Gauss provided a set of mathematical tools to enable astronomers to derive an orbit that provided the best possible fit within the constraints of several imperfect observations. Using his methods, astronomers were able to calculate an accurate orbit far more quickly than they could by trial and error.

Attention soon turned to the question of what the new planet should be called. Inevitably, French astronomers had already started to refer to it as Piazzi, while the Germans, who were keener on mythological names, proposed the names Juno and Hera. Napoleon Bonaparte is even said to have summoned the eminent French mathematician and astronomer Pierre-Simon Laplace to a battlefield conference in order to discuss the naming of the new planet. But Piazzi had determined that he, and he alone, would decide the name.

Piazzi was, by his own admission, a short-tempered, obstinate man, who regarded the compilation of his star catalog as boring but necessary astronomical legwork. He had far greater admiration for the more mathematical astronomers, such as his friend Oriani, whose specialty was orbital calculations. "Your work is genius, mine is muscle," Piazzi wrote to Oriani. "Your study is delightful, and mine is boring. If only I could write such work I would happily renounce the star catalogue."

All of this meant that Piazzi was going to make the most of

his moment of astronomical glory. "I have the full right to name it in the most convenient way to me, like something I own," he declared. "I will always use the name Ceres Ferdinandea, for by giving it another name I will suffer to be reproached for ingratitude to Sicily and the King." His name combined both mythology (Ceres, goddess of the harvest, was also the patron goddess of Sicily) with gratitude to his patron, King Ferdinand of Naples and Sicily. Inevitably, the second half of the name was dropped almost immediately, and the new planet became known as Ceres.

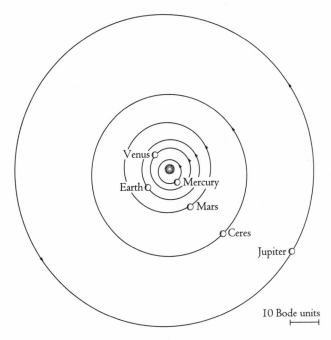

The orbit of Ceres, showing how it fits into the solar system in the gap between Mars and Jupiter. The orbits of Saturn and Uranus are not shown. Planets are not drawn to scale.

Within a few weeks, however, the apparently neat proof of the validity of Bode's law was undermined. On March 28, 1802, another member of the celestial police, Heinrich Wilhelm Olbers, discovered a second planet orbiting the Sun between Mars and Jupiter, and he later named it Pallas. The fact that its orbit was very similar to that of Ceres—the two orbits are tilted with respect to the orbits of the other planets, and overlap—suggested that these two new planets were fundamentally different from the other known planets, none of whose orbits overlapped. And when William Herschel came to measure their diameters, he found that they were absolutely tiny, which suggested to him that they did not deserve to be called planets at all.

Herschel measured the diameters of Ceres and Pallas using a cunning instrument of his own devising called a lamp micrometer. As its name suggests, it consisted of a lamp, but one surrounded by a piece of cardboard punctured with several small circular apertures of different sizes. Assisted by Caroline, Herschel would arrange things so that while he was looking through his telescope (fitted with a very powerful eyepiece) at a planet with one eye, he could see the lamp micrometer with the other. By adjusting the aperture and inserting pieces of colored paper, he was able to make the two images—of the planet and its facsimile—look identical. It was then a simple matter to determine the actual diameter of the planet using trigonometry, given the known size and distance of the lamp micrometer's aperture, and the magnification of the eyepiece. In this way, Herschel determined that Ceres was 162 miles across, and that Pallas was 147 miles across. The modern values are 585 miles and 333 miles, but the point was that both bodies were found to be far smaller than any of the other planets.

As a result, Herschel argued, rather than referring to them as

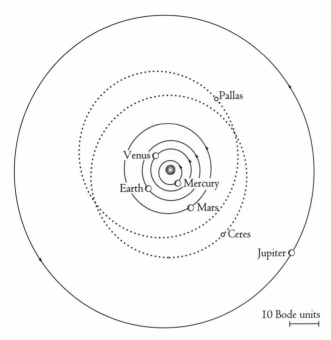

The orbits of the asteroids Ceres and Pallas, which cross one another.

planets, it would make more sense to describe the new bodies as being of a "different species." Initially, Herschel had spent twenty-two fruitless nights searching for Ceres while it had been lost, but had failed to find it because he assumed that it would, like Uranus, appear visibly larger than a star. But Ceres was so small that it had what Herschel called a starlike or "asteroidal" appearance (*astrum* meaning "star" in Latin) in all but the most powerful telescopes. So in May 1802 he sent a paper to the Royal Society in which he suggested that the new bodies should be referred to as "asteroids" rather than planets.

Not everyone liked his idea. Piazzi wrote to Herschel saying that he thought "planetoid" was a better term, since the new objects had nothing to do with stars. Lalande, on the other hand, saw nothing wrong with sticking with the word "planet." But Laplace adopted Herschel's suggestion, as did Olbers himself.

Herschel encountered criticism in the press for his suggestion. "Dr Herschel's passion for coining words and idioms has often struck us as a weakness wholly unworthy of him," declared the *Edinburgh Review.* "The invention of a name is but a poor achievement in him who has discovered whole worlds." Other critics implied that Herschel wanted to distinguish planets from

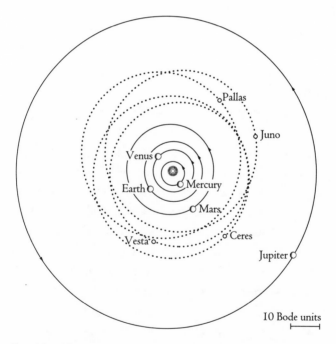

The orbits of the asteroids Ceres, Pallas, Juno, and Vesta.

asteroids in order to make his own discovery seem more important than the findings of Piazzi and Olbers.

Only after the discovery of the third asteroid, Juno—by the German astronomer Karl Ludwig Harding at Lilienthal in 1804—did Herschel respond. He pointed out that "the specific differences between planets and asteroids appear now, by the addition of a third individual of the latter species, to be more fully established; and that circumstance, in my opinion, has added more to the ornament of our system than the discovery of another planet could have done." Unlike planets, asteroids are very small, and their orbits cross one another.

When Olbers discovered the fourth asteroid in 1807, he invited Gauss to name it, in recognition of his valuable contribution to orbital theory. Gauss chose the name Vesta. Olbers also suggested a theory that cleverly salvaged Bode's law: Perhaps the asteroids were fragments of a planet that had previously orbited the Sun at a distance of 28 units but had for some reason exploded. Herschel thought there was some merit in this idea, though he pointed out that more than 30,000 bodies of the size of Pallas would be needed to add up to a body the size of Mercury, the smallest of the planets.

In any case, it had become clear that no planet existed between Mars and Jupiter after all, just a bunch of rocks. Uranus was, however, to show would-be planet hunters the way forward once again.

3

A Very Badly Behaved Planet

That little Vernier, on whose slender lines
The midnight taper trembles as it shines
Tells through the mist where dazzled Mercury burns
And marks the point where Uranus returns.

—SAMUEL PIERPONT LANGLEY
FROM *THE NEW ASTRONOMY*

In the space of a quarter of a century, the map of the solar system had been dramatically redrawn. The discovery of the planet Uranus, orbiting the Sun at twice the distance of Saturn, had doubled the solar system's size; where there had once been a gap between Mars and Jupiter there were now four asteroids. Yet if the five new additions to the family were to take their proper places alongside the already-known planets, it would be necessary to determine their orbits to a comparable degree of precision. To do this, astronomers had two sets of tools: Astronomical instruments were used to measure the positions of celestial bodies, and mathematical methods were then brought to bear to determine their orbits and predict their movements.

Astronomical instruments come in a variety of shapes and sizes, but three pieces of equipment in particular were involved in orbital determination: a transit telescope, a mural quadrant, and an accurate clock. A transit telescope is simply a fixed telescope, pointing due south, and mounted on a tilting bar so that it can pivot up and down but not left and right. Observations are made by noting the exact time that the rotation of the Earth causes a celestial body to cross a vertical wire, or crosshair, strung inside the telescope. The mural quadrant consists of another south-facing telescope that can tilt up and down, this time with a horizontal crosshair, and a metal scale allowing the telescope's angle of inclination to be read off. From a body's time of transit, measured using the transit telescope, and its elevation above the horizon, simultaneously measured by a second observer at the mural quadrant, the body's exact position in the sky can be determined.

The positions of stars and planets are expressed as coordinates on the "celestial sphere," an imaginary grid imposed on the sky by astronomers. Like the surface of the Earth, the celestial sphere has imaginary lines of longitude and latitude ruled on it; its north pole lies directly over the Earth's north pole, its south pole directly below the Earth's south pole, and the celestial equator is the projection of the Earth's equator onto the celestial sphere.

Celestial longitude and latitude, like their terrestrial counter-parts, are measured in degrees, minutes, and seconds; each degree is divided into 60 minutes, and each minute into 60 seconds of arc, so that 1 second of arc, or "arcsecond," is equal to $\frac{1}{3,600}$ of a degree. (Astronomers actually use the terms *declination* and *right ascension*, which are are essentially equivalent to latitude and lon-gitude, respectively.) Distances, as well as positions, in the sky can be expressed in degrees or fractions of a degree. The width of the full Moon, for example, is half a degree, or 30 minutes of arc, or 1,800 arcseconds.

Given the time that a planet passed over the transit telescope's crosshair, its elevation above the horizon, the latitude of the ob-servatory, and the position of the Earth in its orbit around the Sun (derived from the date), astronomers could calculate the planet's coordinates on the celestial sphere. Once these coordi-nates had been established, it was then necessary to carry out a further mathematical correction process called "reduction" be-fore they could be used as the basis of any calculation.

Reduction involves compensating for extremely subtle effects such as precession (the slow wobble of the Earth's axis that causes the position of the celestial north pole to describe a circle over a period of 26,000 years), nutation (a smaller 19-year wobble in the Earth's axis caused by the gravitational effects of the Sun and Moon), and the aberration of light (a tiny shift in the apparent position of heavenly bodies due to the motion of the Earth and the finite speed of light). Since the effects of precession and nuta-tion vary from year to year, astronomers wishing to compare observations made in different years must adjust them to com-pensate for the shifting tilt of the Earth's axis. Only when they have been corrected in this way can observations be used as the raw material for calculations, such as the determination of the orbit of a planet or asteroid.

Modern understanding of planetary orbits goes back to the German astronomer Johannes Kepler. In the early seventeenth century, he analyzed hundreds of planetary observations made by the Danish astronomer Tycho Brahe, and from them derived a set of empirical rules that govern planetary motion. Kepler did not know why his rules worked, but he knew that they allowed the motions of the planets to be predicted with unprecedented accuracy. His most important breakthrough was the "law of ellipses." In 1609, after years of analyzing observations of the planet Mars, Kepler was the first to realize that the shape of its orbit (and, by implication, that of every other planet) was an ellipse, not a circle. Up to that point, astronomers had incorrectly assumed that planetary motion had to be based upon perfect circles.

An ellipse is a regular geometric shape that obeys certain laws. Ellipses can be formed by making a diagonal cut through a cone; depending on the angle of the cut, the shape of the ellipse varies. Some ellipses are almost circular (indeed, a circle is simply a special kind of ellipse), while others have more elongated shapes. It turns out that the orbits of some planets (such as Earth and Venus) are ellipses that are almost circular, whereas the orbits of others (such as Mercury and Mars) are more elongated. Once Kepler had realized that all orbits are elliptical, working out a planet's orbit simply became a matter of finding the particular ellipse that best accorded with the planet's observed positions. (Mathematically, an elliptical orbit can be fully described using six numbers: the planet's average distance from the Sun; a quantity called the eccentricity that describes how far the ellipse deviates from a circle; three angles to specify the orientation of the orbit in space; and the time when the planet is at a particular point on the orbit.)

After an elliptical orbit had been worked out based on a handful of observations, its accuracy could be assessed by checking to see that it agreed with other observations of the planet, adjusting the ellipse if necessary. When an elliptical orbit had been derived that satisfactorily accounted for all previous observations, it could then be used to predict the future position of the planet against the celestial sphere as viewed from Earth. Subsequent observations could be made to check these predicted positions and ensure that the planet was following the expected path through the sky. Any discrepancy could be used to recalculate and refine the orbit further.

Deriving and refining orbits was a tedious, complicated task. But accurate orbits were needed because tables of the future positions of the Moon and planets were used in navigation. For example, having looked up a planet's position on a particular day in a book of tables, a navigator could measure the angle between the planet and the horizon to determine his latitude. More precise astronomical observations of the Moon, or of the moons of Jupiter, made from land during the course of a voyage could also provide an extremely accurate double-check of the accuracy of the chronometers carried by ships to determine longitude.

Calculating, observing, checking, and recalculating thus became the unenviable responsibility of astronomers at official observatories around the world. As observational accuracy and theoretical understanding improved in the seventeenth and eighteenth centuries, the accuracy with which the planets' orbits could be predicted improved correspondingly. The high-quality observations made at the Royal Greenwich Observatory by James Bradley and Nevil Maskelyne were particularly prized by mathematicians, for they formed an unbroken sequence from 1750 onward. (Observations made at other European observatories, in contrast, were prone to interruption by wars and revolutions.)

By the end of the eighteenth century, astronomers could measure and predict planetary positions in the sky to within a couple of arcseconds, or about a thousandth of a degree. The movements of the planets around the Sun were no longer the uncertain wanderings of a few errant stars, as they had been to the ancients; instead, the planets had become the most predictable bodies known to science.

Uranus was expected to slot neatly into this clockwork universe in which observation and theory meshed as neatly as the gear wheels of a chronometer. But instead the new planet threw a monkey wrench into the works. Uranus posed an unexpected challenge to the mathematical theories used to make astronomical predictions. For no matter how elaborate the efforts to define its motion, the wayward planet's position in the sky obstinately refused to match up with the position predicted by theory. And despite all kinds of mathematical gymnastics, nobody could figure out why.

For the astronomers of the late eighteenth century, deriving an accurate orbit for Uranus was particularly tricky because the planet had been discovered so recently. There were only a few years' worth of observations of Uranus, and since it is such a distant and slow-moving planet (taking 84 years to complete one orbit around the Sun), these observations covered only a small arc of its orbit. This made calculations far more error-prone than for the other planets, since a small error in one of the observations could affect the calculation of the orbit quite dramatically.

The scarcity of observational data relating to Uranus made the discovery of Tobias Mayer's 1756 record of the planet in his star catalog very useful—indeed invaluable—to those trying to derive a more accurate orbit. So during the 1780s Johann Elert Bode, who had found Mayer's observation, set about searching through as many star catalogs as he could for other "missing" stars that might actually have been the planet.

Before long, Bode came across a second "ancient" (i.e., pre-discovery) sighting of Uranus: In 1690 John Flamsteed had observed within the constellation of Taurus a faint star whose location roughly coincided with that for Uranus at the time. (By Bode's time, the star was no longer present in the sky.) But its exact position differed slightly from that predicted by any of the orbits that had been worked out, so they were obviously all incorrect. This meant that it would be necessary to calculate an entirely new orbit for Uranus from scratch, taking Flamsteed's 1690 observation into account.

One astronomer who did so was Alexander Fixlmillner of Kremsmünster Observatory in Austria. He derived an orbit using the "ancient" 1690 and 1756 observations, along with a 1781 observation, and one of his own observations made in 1783. He then compared the orbit's predictions with every observation of Uranus he could lay his hands on. He found that these predictions were quite accurate; the largest discrepancy between the predicted and observed positions was just a few arcseconds. Bode

was impressed, and in 1786 he published the details of Fixlmillner's orbit in his astronomical yearbook.

By 1788, however, Fixlmillner's predictions for the position of
Uranus were no longer according with observation. Fixlmillner
tried to adjust his orbit, but he found that any orbit based on
observations made in 1787 and 1788 could not be made to agree
with Flamsteed's 1690 observation. On the assumption the 1690
observation was incorrect, Fixlmillner eliminated it and derived
a new orbit that fit all observations made since 1756 to within 10
arcseconds.

Flamsteed was, however, known to have been an extraordinarily careful observer. This suggested to some astronomers that
there was a more subtle explanation for the incompatibility between the modern observations of the position of Uranus and
Flamsteed's 1690 observation. Perhaps, they thought, it would
only be possible to calculate the orbit precisely by taking into
account the subtle gravitational influences (called perturbations)
of Jupiter and Saturn. This technique, which had been developed
by the French mathematician Pierre-Simon Laplace, had been
used to great effect to improve the accuracy with which the positions of Jupiter and Saturn could be predicted, and was regarded
as the most significant advance in astronomical theory since the
publication in 1687 of the theory of gravitation by the English
physicist and mathematician Sir Isaac Newton.

Newton's insight was that the force that causes an apple to
fall to the ground and the force that holds the Moon in orbit
around the Earth are one and the same. In his epic work *Principia Mathematica*, Newton advanced the law of universal gravitation. According to this law, the gravitational attraction between
two bodies (a planet and the Sun, for example) is proportional to

the product of their masses, divided by the square of the distance between them. One consequence of this law, Newton proved with a clever mathematical argument, was that planetary orbits had to be elliptical, just as Kepler had previously discovered.

During the 1780s, Laplace took Newton's argument one step farther. While the orbit of a single planet around the Sun is elliptical, as Kepler and Newton had shown, Laplace pointed out that in a solar system with multiple planets, the situation is much more complicated. In addition to the gravitational forces between each planet and the Sun, there are much weaker gravitational forces, known as perturbations, between the planets themselves. In most cases, the effect of the planets' gravitational perturbations on each other is so subtle that it can be ignored. But in the case of Jupiter and Saturn, it cannot. They are the two most massive planets, and they move in adjacent orbits; so their mutual perturbations have a noticeable effect on their motions.

Jupiter takes about 12 years to orbit the Sun, and Saturn takes about 30. This means that every 20 years or so, Jupiter overtakes Saturn on its way around the Sun. As the distance between the two planets decreases, the gravitational attraction between them causes Jupiter to speed up very slightly and Saturn to slow down. Once Jupiter has passed Saturn, the perturbatory force acts to slow Jupiter down and speed Saturn up. The perturbation subtly affects the speed and position of each planet, and the shapes of their orbits are slightly distorted, so that they are no longer perfectly elliptical.

Even though it made the mathematics much more difficult, Laplace worked out how to represent the orbits of Jupiter and Saturn as the combination of two motions: a "true" ellipse each

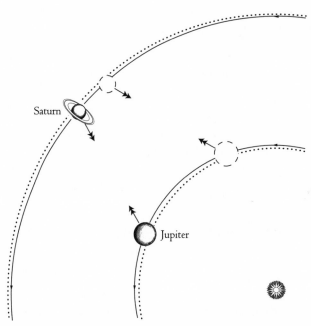

Gravitational attraction between Jupiter and Saturn distorts their orbits away from the "true" elliptical paths (shown dotted) that they would otherwise follow around the Sun, and slightly affects each planet's position. By taking these subtle perturbations into account, astronomers can predict the positions of the two planets far more accurately.

planet would describe if it was the only planet orbiting the Sun, plus the perturbing motion due to the other planet. With these new corrections taken into account, the motions of Jupiter and Saturn could be predicted far more accurately than ever before. This work earned Laplace the illustrious nickname "the Newton of France."

The same approach was now applied to the orbit of Uranus,

and in 1790 the French Academy of Sciences awarded a prize to Jean-Baptiste-Joseph Delambre for deriving an orbit for Uranus that took into account the gravitational effects of Jupiter and Saturn. This involved subtracting the effects of Jupiter and Saturn from each observed position of Uranus, to determine where it would have been had the other planets not been present and work out the true elliptical orbit it would have followed in their absence. Finally, the effects of Jupiter and Saturn were added back in, to give an accurate representation of the actual future motion of Uranus. Delambre's orbit was based on the ancient observations made by Flamsteed in 1690 and Mayer in 1756, plus two more made by Pierre Charles Le Monnier, a gifted but unlucky French astronomer who realized in 1788 that he had observed Uranus on two occasions in the 1760s but had mistaken it for a star.

The tables based on Delambre's calculations were published in 1791. At last—by taking into account the gravitational effects of the other major planets—the orbit of Uranus appeared to have been pinned down. In Oxford, Thomas Hornsby observed Uranus in 1798 and found that its position agreed with Delambre's predictions very closely, to within a few arcseconds. The errant planet had, it seemed, been tamed.

Uranus's period of good behavior did not, however, last long. After 1800, discrepancies started to become apparent between the planet's position as predicted by Delambre and as measured in the sky. Evidently its orbit would, once again, have to be totally recalculated.

The next astronomer to tackle the problem was Alexis Bouvard, a mathematical prodigy who had begun his astronomical career observing the stars at night as a shepherd boy. He then attended a series of public lectures on mathematics in Paris and rapidly made a name for himself as a skilled mathematician. Bouvard was employed as a "computer" (a mathematical assistant) by Laplace, was elected to the Academy of Sciences, and won a prize for his work on the orbit of the Moon.

In 1808, Bouvard published a set of tables of the positions of Jupiter and Saturn, which subsequently proved to be inaccurate, due to the incorrect values of the planetary masses on which they were based. Bouvard decided to derive a set of new, more accurate tables. At the same time, he planned to sort out, once and for all, the problem of Uranus. He started by mounting his own comprehensive search for accidental observations of the planet prior to its discovery by Herschel.

In particular, Bouvard went back to the records of Le Monnier. Bouvard's close examination of Le Monnier's records revealed that the unfortunate astronomer had actually recorded Uranus a total of twelve times, including the two observations he had subsequently discovered himself. "Through not comparing his observations from day to day, Le Monnier was deprived of the honour of a beautiful discovery," noted Bouvard. He also complained about the untidiness of Le Monnier's records. On one occasion, Bouvard claimed, Le Monnier had noted the details of one of his observations of Uranus on the back of a perfumier's paper bag that had previously contained hair powder.

Meanwhile, a similar search of Flamsteed's records was being carried out by Johann Karl Burckhardt, who found that Flamsteed had recorded Uranus as a star in 1712 and in 1715, while observing the planet Saturn. These observations were important because they plugged the huge gap in the known observations of Uranus between 1690 and 1750. Friedrich Wilhelm Bessel, director of the Königsberg Observatory in Prussia, uncovered the next observation: The English astronomer James Bradley had recorded the position of Uranus in 1753.

Armed with more ancient observations of Uranus than anyone before him, Bouvard started work on a new orbit in 1820. It quickly became clear to him that something very strange was going on; it was simply not possible to find an orbit that accorded with all the observations, even when perturbations of Jupiter and Saturn were taken into account. The errors were too large—as much as 120 arcseconds in places. "Hence," declared Bouvard in his introduction to the tables, which were published the following year, "it is the exactness of the ancient observations where the doubt falls. And this is difficult to avoid when considering the circumstances in which they were made."

Bouvard decided to dodge the problem by declaring all the ancient (i.e., pre-1781) observations to be inaccurate and untrustworthy, and brushing them under the carpet. He pointed out that Bradley's and Mayer's observations were single observations; that Flamsteed's instruments were suspected of being inaccurately calibrated; and that Le Monnier was generally not to be trusted. (The story about the paper bag was probably not true and may have been circulated by Bouvard in order to discredit Le Monnier's observations. While Le Monnier had recorded some observations on a paper bag, they were not of Uranus.) This would also explain why Delambre's tables, which had been drawn up using both ancient and modern observations, had failed to explain the motion of Uranus.

"As a result of these considerations I have suppressed the ancient observations and based the new tables on modern ones only. The agreement shown is good," Bouvard concluded. But, perhaps aware of the thin mathematical ice on which he was standing, he decided to cover himself both ways, adding: "I leave to the future the task of discovering whether the difficulty of reconciling the two systems results from the inaccuracy of the ancient observations, or whether it depends on some extraneous and unknown influence which may have acted on the planet."

In fact, the discrepancies between the historical positions predicted by Bouvard's new orbit and the observations of Flamsteed, Mayer, and Le Monnier were huge: up to 60 arcseconds, far too large to be attributed to observational error alone. Bouvard's suggestion that such eminent astronomers had consistently made enormous errors in their observations bordered on the slanderous; fortunately for him, all the astronomers in question were dead. Furthermore, even some of the modern observations differed by 10 arcseconds from his new orbit; again, by the standards of the day, this was a suspicious disparity.

The new tables were clearly a fudge and were immediately criticized by Bessel, who pointed out that Bouvard also seemed to have made a number of mathematical errors. Another objection to Bouvard's tables was that his estimate of the mass of Uranus, made on the basis of its perturbatory effects on Saturn, was also unusually high. But for the most part, astronomers were happy with Bouvard's tables, because as new observations of Uranus were made, they accorded well with his predictions. Like Delambre three decades before him, Bouvard appeared to have subdued the unruly planet.

And so things stood at the end of William Herschel's extraordinary astronomical career. Having started out as an amateur,

Herschel had become renowned as a telescope maker and the discoverer of an entirely new planet and two accompanying moons; he had identified over 1,000 double stars; he had discovered two new moons of Saturn and determined the orbital period of Saturn's rings; and he had been the first to notice seasonal changes in the ice caps of Mars. He died on August 25, 1822, surrounded by his astronomical papers and charts, and with his sister Caroline, as ever, at his side. By a strange coincidence, he had lived for eighty-four years—which just happens to be the length of time his planet takes to complete an orbit around the Sun.

By the time of William Herschel's death his son, John, was fast establishing himself as an eminent scientist in his own right. In fact, John Herschel was to play a key role in the subsequent unraveling of the mystery of Uranus. For it soon became apparent that, once again, all was not well with his father's notoriously badly behaved planet.

Problems resurfaced in 1825 and 1826, when observations of Uranus made in Austria showed small but definite discrepancies from the positions predicted by Bouvard's tables. By 1828, further observations made in England under the supervision of George Biddell Airy, professor of astronomy at Cambridge University, showed errors of 12 arcseconds. As one astronomer pointed out at the time, the tables for the other planets were not quite perfect, and the errors varied slightly; but in the case of Uranus the error just got bigger and bigger. The error in the planet's longitude rose to 16 arcseconds; by 1829 it had reached 23 arcseconds, and by 1830 it was 30 arcseconds.

This error was simply too large to ignore, as Airy pointed out in his *Report on the Progress of Astronomy*, written in 1832 for the

British Association for the Advancement of Science—the report that John Couch Adams would come across, nine years later, in a Cambridge bookshop. With respect to Uranus, wrote Airy, "a singular difficulty occurs. . . . it appears impossible to unite all the observations in one elliptical orbit, and Bouvard has rejected the ancient ones entirely. But even thus the planet's path cannot be represented truly."

Clearly, something funny was going on. The victim was Uranus; the crime, orbital interference; but the identity of the culprit was a complete mystery.

4

An Astronomical Mystery

No marvel, indeed, that Uranus has come to be accounted the puzzle of our science—no wonder that so many minds were turned to this portion of the celestial mechanism.

> —J. P. NICOL, PROFESSOR OF ASTRONOMY AT THE
> UNIVERSITY OF GLASGOW, 1848
> FROM *THE PLANET NEPTUNE, AN EXPOSITION
> AND HISTORY*

By the 1830s the strange behavior of Uranus had been going on for decades, so astronomers had had plenty of time to devise a variety of explanations to account for it. Some of these theories, however, were decidedly more credible than others.

One theory, which had made the rounds when Alexis Bouvard's tables were first published in 1821, was that Uranus might have been hit by a comet. This would have neatly explained the apparent lack of agreement between the pre-1781 and post-1781 observations. The comet, which would have to have collided with the planet sometime between 1771 (when it was last seen by Le Monnier) and 1781 (when it was first seen by William Herschel), could have nudged the planet, causing it to deviate from the orbit it would otherwise have followed.

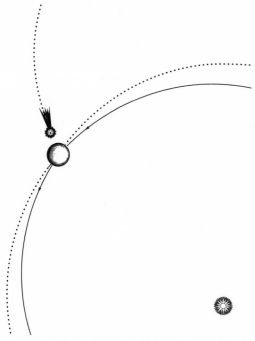

The cometary impact theory. Some astronomers suggested that an enormous comet might have struck Uranus between 1771 and 1781, thus altering the planet's orbit.

While this theory would have neatly justified Bouvard's decision to discard the old observations, it would only have held water if his 1821 tables, based on the post-1781 observations, had continued to predict the position of Uranus accurately. Unfortunately, after 1825 or so, they didn't. Short of postulating another unseen cometary impact in 1825, there was no way to reconcile the idea of a cometary impact with Uranus's behavior. So this theory was swiftly discarded.

Another possible explanation involved the presence of a resistive medium in space. Since, from the point of view of the position predicted by the Bouvard tables, Uranus was falling behind schedule, perhaps something was slowing it down. There was a precedent for this suggestion: The idea of a "thin ethereal medium" had originally been proposed by the German astronomer Johann Franz Encke to explain the anomalous motion of a comet.

In 1819 Encke realized that the comets observed in 1786, 1795, 1805, and 1818 were actually one and the same. (The 1795 discovery was, in fact, made by Caroline Herschel.) He worked out an elliptical orbit for the comet and found that it went around the Sun once every 3.3 years. But as he refined the comet's orbit following its reappearances in 1822 and 1829, Encke found that even when planetary perturbations were taken into account, each revolution around the Sun was taking about 2.5 hours less than the last, so that the period of the comet's orbit had decreased by nearly 2 days since 1786. Encke performed an elaborate calculation in which he showed that this phenomenon could be explained by assuming the presence of a resistive medium, the effect of which was to slow the comet down and cause its orbit to become smaller and tighter, so that each revolution took less time than the last.

But could this resistive medium, if it existed, explain the behavior of Uranus? It seemed unlikely. For some reason, the effects of this resistive medium were apparent only in the case of
Encke's comet; with the exception of Uranus, all the other planets, asteroids, moons, and comets behaved just as Newton's gravitational theory predicted. The idea of a resistive medium that
affected only two objects in the entire solar system was rejected
as absurd. (The strange motion of Encke's comet is now believed
to be the result of the asymmetrical ejection of dust and gas from
the comet, which provides a very gentle but uneven thrust.)

If the cometary impact and resistive medium theories
sounded outlandish, a third candidate—the idea that Uranus's
motion might be being influenced by an unseen moon orbiting
the planet—was even less plausible. Only an absolutely enormous moon, comparable in size to Uranus itself, would be able
to affect the planet's motion; and while Uranus was known to
have two tiny moons, no such large moon had ever been seen.
Furthermore, the effect of a massive satellite, even if one existed,
would be to cause a small periodic wobble in the position of
Uranus, rather than a constantly increasing error in one direction. So this theory was also rejected.

Another possibility was that Bouvard had made some kind of
mathematical error in drawing up his tables. No astronomical
tables were entirely error-free, and some were riddled with mis-
takes. Indeed, since mathematical tables (of logarithms, for example) were used in the process of compiling astronomical
tables, errors could propagate from one set of tables to another.
In the words of John Herschel, "An undetected error in a logarithmic table is like a sunken rock at sea yet undiscovered, upon
which it is impossible to say what wrecks may have taken place."
Such errors could not, however, be blamed for the discrepan-

cies between the predicted and observed positions of Uranus. When Bouvard's tables were examined very carefully by a number of mathematicians, some minor arithmetical errors were found. His estimate of the mass of Jupiter, which was crucial in calculating the effects of perturbations on Uranus, was shown to be inaccurate, and a set of corrections was introduced to enable users of the tables to compensate for this mistake. But even when these adjustments were taken into account, the tables still failed to explain the motion of Uranus. Clearly, something more than simple miscalculation was to blame.

Another possible explanation, though it had few adherents, was that Newton's law of gravitation was incomplete in some way. Perhaps Newton's law did not apply to Uranus, since it was so far from the Sun. Or perhaps the force of gravity might affect different bodies in different ways, depending on their chemical composition.

The problem with both of these ideas was that Newton's law of gravitation was the finest example of a scientific law in existence. On previous occasions when it had been called into doubt, it had always been subsequently vindicated. It seemed that the law was truly universal; there could be no exception made for Uranus. As a result, most astronomers regarded the prospect of tinkering with the law of gravity as a last resort, to be contemplated only when all other possibilities had been eliminated.

This left just one explanation. It was known that Jupiter, Saturn, and Uranus had sizable effects on each other's orbits as a result of gravitational perturbations. Might there be another large, undiscovered planet, even more distant from the Sun, whose perturbations were affecting the motion of Uranus? The more William Herschel's planet refused to behave itself, the more people were inclined to consider this intriguing prospect.

The idea that there might be another planet orbiting beyond Uranus was not new. Vague speculation about a more distant planet had been going on ever since the discovery of Uranus, and the apparent proof of the validity of Bode's law provided by the discovery of the asteroids seemed to offer a clue to its likely distance from the Sun. According to Bode's law, the new planet, if it existed, would have an orbit with an average radius of 388 units, or about twice Uranus's distance from the Sun. (See box below.) In 1802 one astronomer, Ludwig Wilhelm Gilbert, even went so far as to propose a name for this hypothetical planet: "Is Ophion, a planet beyond the orbit of Uranus, another undiscovered world?"

The strange behavior of Uranus prompted renewed speculation during the 1830s. In 1835 Jean Valz, director of the Marseilles Observatory, wrote to François Arago, the leading French astronomer, on the subject of Halley's comet, which was due to return that year and had not quite been following the expected orbit. Since the comet's orbit carried it beyond the orbit of Uranus,

Planet	Radius of Orbit	Predicted Radius
Mercury	4	$4 + 0 = 4$
Venus	7	$4 + (1 \times 3) = 7$
Earth	10	$4 + (2 \times 3) = 10$
Mars	15	$4 + (4 \times 3) = 16$
(asteroids)	28	$4 + (8 \times 3) = 28$
Jupiter	52	$4 + (16 \times 3) = 52$
Saturn	95	$4 + (32 \times 3) = 100$
Uranus	192	$4 + (64 \times 3) = 196$
?		$4 + (128 \times 3) = 388$

Valz thought that he could explain these orbital discrepancies by postulating the existence of "an invisible planet, located beyond Uranus." Such a planet could also, he suggested, explain the anomalous motion of Uranus. "Would it not be admirable to ascertain the existence of a body which we cannot even observe?" he suggested.

In 1834 George Airy, director of the Cambridge Observatory, whose report on "progress in astronomy" had shown him to be particularly interested in the question of Uranus, received a letter making a similar suggestion. Thomas Hussey, an English clergyman who was also a keen amateur astronomer, wrote that "the apparently inexplicable discrepancies between the ancient and modern observations suggested to me the possibility of some disturbing body beyond Uranus, not taken into account because unknown."

Hussey explained to Airy that he had considered trying to work out the approximate position of the planet mathematically and then mapping a small area of the sky to look for any moving stars over the period of a week, to see if he could find the planet. But, he lamented, "I found myself totally inadequate to the former part of the task." Hussey had, he added, recently met Bouvard in Paris and had asked him what he thought of this idea. Bouvard said he had thought of it too, and that he was considering doing the calculations himself. Airy was known to be a skilled mathematician; what, asked Hussey, did he think?

Airy poured cold water on Hussey's scheme. "I have often thought of the irregularity of Uranus, and since the receipt of your letter I have looked into it more carefully," he wrote. "It is a puzzling subject, but I give it, as my opinion, without hesitation, that it is not yet in such a state as to give the smallest hope of making out the nature of any external action on the planet." Airy noted that Bouvard's tables gave similar errors for the position of the planet in 1750 and 1834, when the planet was in the same position relative to the Sun; since any perturbing planet would have moved between these two dates, its influence would be expected to result in a variation of the error. Airy concluded that this didn't appear to be an irregular perturbation and expressed his view that further refinement of the orbit, including a more detailed recalculation of the perturbations due to Saturn, would solve the problem. He said that he hoped to perform such calculations, which he had already sketched out in rough form, when he had the time.

Finally, Airy said that even if Uranus was subject to irregular perturbation, "I doubt very much the possibility of determining the place of the planet which produced it. I am sure it could not be done till the nature of the irregularity was well determined from several successive revolutions." In other words, Airy believed that Uranus would have to complete several orbits around the Sun before the action of any perturbing planet would become apparent. Since Uranus takes 84 years to complete one orbit, this would mean waiting for hundreds of years.

Airy had in fact been keeping a keen eye on Uranus, and he made several observations of the planet between 1833 and 1835. His analysis of these observations, published in the German journal *Astronomische Nachrichten* (Astronomical News) in 1838, showed that Bouvard's tables were not merely inaccurate in their predictions of the planet's longitude; their predictions of its distance from the Sun, a quantity known as the radius vector, were wrong too.

In 1837 Airy received a letter from Alexis Bouvard's nephew

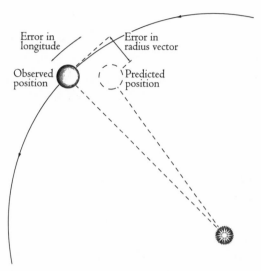

The error in the radius vector. Bouvard's tables predicted the longitude and the radius vector of Uranus. While most astronomers were primarily concerned with the errors in longitude, George Airy showed that Bouvard's predictions of the radius vector were wrong too.

Eugène, who wrote that he was "working on a task which I believe is not without importance." He explained that his uncle was working on new tables for Jupiter and Saturn, and that it had fallen to Eugène to prepare new tables for Uranus to accompany them. Everyone knew, he said, that the existing tables for Uranus were in error and getting worse all the time. "Does this suggest," he asked Airy, "an unknown perturbation, arising in the motions of this planet, by a body situated further away? I don't know, but that is my uncle's idea, at least. I regard the solution of this question as of the utmost importance. But to answer it, I need to reduce the observations of Uranus with the utmost precision, and in many respects I lack the means to do so."

In his reply, Airy repeated what he had suggested to Hussey, namely, that in his opinion the errors could be fixed by recalculation of the effects of Jupiter and Saturn. The errors in the longitude of Uranus were, he said, increasing "with fearful rapidity. . . . I cannot conjecture the cause of these errors, but I am inclined, in the first instance, to ascribe them to some error in the perturbations. If it be the effect of some unseen body, it will be nearly impossible ever to find out its place."

Why was Airy so skeptical about the idea of an undiscovered planet? Such skepticism was, arguably, entirely in keeping with his character. Airy was a methodical, unimaginative man who had been appointed astronomer royal in 1835 in order to restore the reputation of the Royal Greenwich Observatory, which had become somewhat tarnished under his predecessor, John Pond. Pond is today best remembered for introducing the red ball on the roof of the observatory that rises and falls at one o'clock to act as a time signal to shipping moored nearby on the Thames; he also instituted a thorough reform of the observatory, but clung to his post despite ill health and allowed standards to slip. Having run the Cambridge Observatory for many years, Airy had a reputation as a good organizer and as possessing a safe pair of hands. He was, however, very set in his ways and seems to have been a rather difficult character, judging by the number of anecdotes that detail his strange behavior.

Airy insisted, for example, that observers remain on duty at the observatory in cloudy or rainy weather, even though no observations could be made; he would walk between the instruments ensuring that every man was at his station. He was just as strict when skies were clear. On one occasion he came across one of the observatory's assistants observing on his day off. Airy asked the man what he was doing; "looking for new planets" was the reply. Airy reprimanded the man and sent him home at once. As far as Airy was concerned, such speculative work, even when conducted after hours, was not the purpose of the observatory. "The observatory was expressly built for the aid of astronomy and navigation, for promoting methods of determining longitude at sea, and more especially for determining the moon's motions," he declared. Anything outside these clearly defined tasks—such as looking for supposed new planets—was to be frowned upon as an unnecessary diversion.

Another tale tells of Airy spending a whole afternoon writing the word "empty" on large pieces of cardboard, which were then nailed to empty packing crates so that they could be distinguished from other boxes. Airy chose to perform this menial task himself because asking another member of the staff to do it would have upset the observatory's strictly time-tabled regime.

Even Airy's friends and family thought his behavior was rather odd at times. One of his friends said that "if Airy wiped his pen on a piece of blotting paper he would duly endorse the blotting-paper with the date and particulars of its use, and file it away amongst his papers." Airy's son recalled that "his accounts were perfectly kept by double-entry throughout his life, and he valued extremely the order of book-keeping. . . . he seems not to have destroyed a document of any kind whatever: counterfoils of old cheque-books, notes for tradesmen, circulars, bills and correspondence of all sorts were carefully preserved in the most com-

George Biddell Airy (*The Great Astronomers* by Robert S. Ball, London: Sir Isaac Pitman and Sons, 1895)

plete order, and a huge mass they formed." Indeed, at one point Airy explicitly issued a "general order" to his staff that "no paper whatever is to be destroyed. It is to be delivered to me, or to be lodged in the portfolio or other place prepared for its preservation."

Without doubt, Airy ran a tight ship. He standardized and streamlined procedures at the observatory, turning its analytical operations into a mathematical production line. He introduced preprinted forms to assist with reduction calculations, and set his team of mathematicians to work clearing the backlog of unreduced observations left behind by his predecessors. Walter

Maunder, one of his assistants, noted that "his regulation of his subordinates was despotic in the extreme—which was the cause of not a little serious suffering to some of his staff, whom, at the time, he looked upon as mere mechanical 'drudges'. The unfortunate boys who carried out the computations of the great lunar reductions were kept at their desks from eight in the morning till eight at night, without the slightest intermission, except an hour at midday."

Perhaps the most striking illustration of Airy's bureaucratic attitude toward his work is the fact that after being appointed astronomer royal, he hardly ever looked through a telescope again. Indeed, he had terrible eyesight and commonly carried several pairs of glasses with him, some of which he had designed himself. During his reign at Greenwich he probably made fewer observations than any previous astronomer royal; between 1835 and 1843, for example, he made 164 of the 69,204 observations recorded at the observatory. Airy regarded his task as one of administration, not observation.

His reputation as a solidly reliable man of science meant that Airy was called upon by the British government to rule on a number of pressing scientific matters. He was appointed to the Railway Gauge Commission, a body set up to decide on the appropriate gauge for Britain's railways; he advised on the construction and operation of the Westminster Clock, known as Big Ben; he helped with the reestablishment of British weights and measures following a fire at the Houses of Parliament; and he advised the Royal Navy on the use of magnetic compasses in iron ships. In 1842 Airy was asked to determine whether or not the mechanical computer designed by the mathematician Charles Babbage, construction of which had started several years earlier, should receive further government funding. Airy pronounced Babbage's scheme "worthless," and that was the end of it; the unassembled parts were melted down for scrap.

Airy had been appointed astronomer royal because he was dependable; he was not the sort of man to take a leap into the scientific unknown. So it is not surprising that Airy was so unreceptive toward the provocative notion of a new planet orbiting beyond Uranus. Increasingly, however, he was isolated in his belief that it would be possible to explain the motion of Uranus without one.

One of the strongest proponents of the idea of a new planet was Friedrich Wilhelm Bessel, director of the Königsberg Observatory in Prussia. Bessel had previously been a proponent of the idea that the force of gravity might vary for bodies of different chemical compositions, but after carrying out a series of experiments to investigate this theory he decided it was incorrect. He then adopted the new planet theory with the evangelical zeal of a convert.

In 1840 Bessel gave a lecture in which he said that any attempt to explain the behavior of Uranus would have to be "based on the endeavour to discover an orbit and a mass for some unknown planet, of such a nature that the resulting perturbations of Uranus may reconcile the present lack of harmony in the observations." He was convinced that years of dogged mathematical effort would eventually reveal the unknown planet, but that it would be hard work: "The vein of pure gold lies deep; we must continuously and furiously dig for it . . . because at this time the worker deep below sweats only for rock and poor earth. The smelting of this is a meagre reward before the finding of the vein." Bessel and one of his pupils, Friedrich Flemming, started laying the mathematical groundwork for such a calculation with a reanalysis of the historical observations of Uranus. But Flemming died unexpectedly, and Bessel subsequently fell ill himself, so nothing came of their attempt.

Another astronomer who publicly endorsed the idea of an unseen planet was Johann von Mädler, director of the Dorpat Observatory in Estonia. In his book *Popular Astronomy*, published in 1841, he explained that the most likely solution to the mystery was "a planet acting upon and disturbing Uranus; we may express the hope that analysis will at a future time realise in this her highest triumph, a discovery made with the mind's eye, in regions where sight itself is unable to penetrate."

That very year, the same thought occurred to John Couch Adams as he stood in a Cambridge bookshop, reading Airy's report on "progress in astronomy." But Adams, unlike those who had gone before him, had both the opportunity and the mathematical skills necessary to take the next step.

5

The Young Detective

Nature that framed us of four elements,
Warring within our breasts for regiment,
Doth teach us all to have aspiring minds:
Our souls, whose faculties can comprehend
The wondrous architecture of the world,
And measure every wandering planet's course . . .

—CHRISTOPHER MARLOWE
FROM *TAMBURLAINE THE GREAT*

$\mathbf{I}$t is easy to see why the challenge of determining the position of an unseen planet from nothing more than a sheet of numbers appealed so strongly to John Couch Adams: He had been interested in both mathematical problems and astronomy from a very early age. Born in a farmhouse in Lidcot, Cornwall, in 1819, Adams was the oldest of seven children. His mathematical talent became evident early in his childhood, and by the age of eleven he had acquired a reputation as something of a prodigy. Adams preferred to keep his mathematical ability to himself, but his father was so proud of his son that he liked to show him off in public.

In 1830 Adams and his father went to visit Mr. Pearse, a family friend in Devon who had a son of similar age to John. The two men got into an argument over which of their sons was the more talented mathematician, and decided to resolve it with a problem-solving contest. Each boy was asked to propose a mathematical problem for the other to solve. Adams was unable to solve young Pearse's problem and claimed it was insoluble. Pearse could not solve Adams's problem either, which was to "find the sum of money such that when divided in two, the pounds and shillings are reversed."

Since neither boy could solve the other's problem, each was asked to solve his own. Pearse admitted that his problem was insoluble; Adams, on the other hand, gave the answer to his problem, which is £13 6s. (Since there are twenty shillings in an old English pound, half of this amount is £6 13s.) Adams was declared the winner, but Mr. Pearse proposed another challenge, this time against the local schoolmaster. George Adams, John's brother, later recalled that John "again distinguished himself, and not only gave a proof of his proficiency beyond that of his rival Pearse, but the master also, who set an equation that he himself could not solve, but John did so, to the astonishment of his friends."

Soon afterward, Adams's mathematical interests took an astronomical turn. By the age of fourteen he was drawing his own star maps and reading all the books on mathematics and astronomy he could lay his hands on. In 1834 he was given a copy of Sir John Herschel's book *Astronomy* as a school prize. (By this time John Herschel, following in his father's footsteps, had established himself as one of Britain's leading astronomers and was knighted in 1833 by King William IV.) The following year Adams saw a comet for the first time and became interested in calculating orbits and working out the times of eclipses. At the age of sixteen he predicted the exact time a forthcoming eclipse of the Sun would be visible from Lidcot—an extremely complex calculation. He also made his own astronomical measurements, making marks on a windowsill in order to record the positions of noonday shadows, and created an instrument out of cardboard for measuring the elevation of the Sun.

Adams was almost entirely self-taught. He learned his mathematics from books and liked to derive his own methods and proofs. He often preferred to work something out for himself rather than simply accept someone else's explanation. He was, for example, a good singer and violinist, but his understanding of musical theory was founded on mathematics, as a diary entry from his youth attests: "This afternoon by the application of mathematical reasoning I discovered for the first time the relative lengths of the vibrations which form the different notes and semi-tones, and consequently penetrated the mysteries of flats and sharps being attached to keys, which had often puzzled me before." He was also aware of the importance of perseverance when trying to solve a difficult problem. As he remarked to his brother Thomas, who was having difficulty with algebra, "the harvest does not immediately follow the sowing."

In 1839 Adams went to study with a local curate who had previously studied mathematics at Cambridge University. In October he took the university's entrance exams and won a place at St. John's College as a "poor sizar." This meant that his college fees were slightly reduced in return for doing teaching work for the college. Even so, sending John to Cambridge was the cause of considerable financial hardship for his family.

Adams flourished at the university. As well as attending five hours of lectures each day, he enjoyed singing and playing cards, though his mind tended to wander if he was grappling with a mathematical problem. "You could sometimes see in his eyes that his mind was far away and then you did not wonder at his breaking the simplest rules of the game," said a friend with whom Adams played whist. Adams's working method involved thinking through a problem before setting pen to paper. His cleaning woman recalled that she often found him lying on his sofa, deep in thought, with no books or papers anywhere near him. Adams liked to spend Sunday evenings with a group of friends at Trinity College in the rooms of a don known as Carus, who was a charismatic and influential figure. These visits may have inspired Adams in another way: Carus's rooms had formerly been the lodgings of Isaac Newton, who had written his greatest work, the *Principia Mathematica*, within its walls.

While studying, Adams did his best not to spend too much time on astronomical diversions that had nothing to do with his degree. He did not always succeed, as one diary entry from early 1841 shows: "I have badly broken my plan today, chiefly wasting my time with astronomy. I resolve not to let my astronomical amusements interfere with my regular work." At Easter that year, Adams visited the university observatory and saw its magnificent telescope, the Northumberland equatorial, which was one of the finest telescopes in the country.

And on June 26 the course of Adams's future career was determined when he came across Airy's report on "progress in astronomy" in the local bookshop, read about the misbehavior of Uranus, and realized that here was a mathematical and astronomical challenge that he could really sink his teeth into.

From that day forward, Adams made no secret of his determination to solve the mystery. When a fellow student asked him about his future plans, Adams replied: "You see, Uranus is a long way out of his course. I mean to find out why. I think I know."

John Couch Adams (*The Great Astronomers* by Robert S. Ball, London: Sir Isaac Pitman and Sons, 1895)

It was, however, another two years before Adams could devote his full attention to the problem.

In 1843 he had to take his final graduation exams, the dreaded Mathematical Tripos, consisting of twelve papers of three hours each, and so named because of the three-legged stool on which the examiner traditionally sat. The Tripos was a daunting challenge to even the finest students; its end result was a list, ranking the entire year of mathematics students in order of merit, from the student with the highest score (known as the "Senior Wrangler") to the student with the lowest (the "Wooden Spoon").

Despite his unassuming demeanor, Adams was quickly identified as a likely candidate for Senior Wrangler. One of his contemporaries at St. John's noted that Adams was "rather a small man, who walked quickly, and wore a faded coat of dark green," but who was, nonetheless, respected for his mathematical skills. "The fastest and vainest man would have been civil to Adams, for he was known to be a pretty certain Senior Wrangler . . . men bet on him and backed him as they would a racehorse." Adams himself was characteristically modest about his chances; he assured Ballard, the college porter, that it was not at all certain that he would emerge as Senior Wrangler, and discouraged Ballard from betting too heavily on him.

As it turned out, Adams trounced the other mathematicians in his year. One of them observed that "in the Tripos examination I noticed that when everyone was writing hard, Adams spent the first hour in looking over the questions, scarcely putting pen to paper the while. After that he wrote out rapidly the problems he had already solved in his head and ended by practically flooring the examiners." Adams scored over 4,000 marks, more than twice as many as the next best student, and was duly named

Senior Wrangler. A few weeks later he was awarded the First Smith's prize, the university's most esteemed mathematical trophy. Having proved himself academically, Adams was finally in a position to turn his attention to his own personal quest: the search for the explanation for the misbehavior of Uranus.

By the time Adams got to work on the problem, back at home in Lidcot in the summer of 1843, astronomers had long since given up on the idea that Bouvard's tables of the position of Uranus bore any semblance to reality. In 1837 Airy had referred to the "fearful" discrepancy between the planet's longitude as predicted in the tables and observed in the sky. And this discrepancy continued to grow: The error increased from a totally unacceptable 30 arcseconds in the early 1830s, to 50 arcseconds by 1838, and 70 arcseconds in 1841. And, of course, Airy had shown that the predicted values of the planet's radius vector were wrong too.

There was, in other words, no question that Bouvard's tables were flawed. But before proceeding any farther, Adams had to eliminate the possibility that the tables were wrong because of an error in the mathematical argument Bouvard had used to construct them. If Bouvard had made a mistake when applying the theory of perturbations, for example, that would also explain why his tables failed to correspond with observations. So Adams examined the orbit of Uranus in far greater detail than Bouvard had, including extra mathematical terms that Bouvard had neglected as being too insignificant to bother with. Adams also took into account the fact that Jupiter's mass was known more accurately than it had been twenty years earlier, when Bouvard had drawn up the tables. Despite all this, he found that the motion of Uranus was still definitely anomalous, even when compared with his slightly corrected version of Bouvard's tables.

The time had come to embark on the calculation that had so far defeated every mathematician who had attempted it: to determine the position and orbital characteristics of an unseen planet that could account for the anomalous motion of Uranus. This was, in essence, a problem of inverse perturbation. Astronomers knew how to calculate the deviation of a body (Saturn, say) from its orbit under the influence of a known perturbing body (Jupiter, for example). Since Jupiter's mass and orbit were known, its gravitational effects on Saturn could be calculated relatively easily. They could then be subtracted from Saturn's observed motion to reveal its unperturbed elliptical orbit—the elliptical path it would follow around the Sun if Jupiter did not exist. Having established this "true" ellipse as the foundation of the orbit, the perturbations due to Jupiter could then be taken into account as well, enabling Saturn's position to be predicted with extraordinary accuracy.

But in the case of Uranus, the problem was the other way round. If Adams was correct, it was being perturbed by Jupiter and Saturn, whose masses and orbits were known, and also by an unseen planet, whose characteristics were not. So it was impossible to subtract the perturbations from its observed motion and derive the true ellipse. Calculating the true ellipse depended on knowing the perturbations due to the unseen planet; but those perturbations could only be calculated if the true ellipse was known. Faced with this mathematical chicken-and-egg situation, Adams had no choice but to solve both problems simultaneously.

He decided to start off by attacking a simplified version of
the problem: He would assume that the unseen planet had a
perfectly circular orbit, at the distance from the Sun predicted by
Bode's law, namely twice the average distance of Uranus. These
two assumptions made the mathematics simpler and would enable Adams to check that he was barking up the right mathematical tree. His plan was to conduct a mathematical experiment, to
calculate how the presence of this theoretical planet would affect
the motion of Uranus. Would it produce the kind of erratic motion that had actually been observed over the years by astronomers? If it did, Adams could go on to refine his calculation and
work out the exact characteristics of the planet responsible.

John's brother George helped out, by watching John as he
worked to ensure he did not make any careless arithmetical errors. In an account titled "Reminiscences of Our Family," published many years later, George recalled: "Frequently, night after
night, I have sat up with him in our little parlour at Lidcot when
all else had gone to bed, looking over his shoulder seeing that he
copied, added and subtracted his figures correctly to save his
doing it twice over. On those occasions dear Mother, who would
be exhausted with her heavy work, before going to bed would
prepare the milk and bread for us for supper before retiring. This
I should warm when required and we take together. Often have
I been tired, and said 'It's time to go to bed, John'. His reply
would be, 'In a minute,' and go on almost unconscious of anything but his calculations. In his walks on those occasions on
Laneast Down, often with me, his mind would be fully occupied
in his work. I might call attention to some object, and get a
reply, but he would again relapse into his calculations."

As he worked, Adams knew he was venturing into uncharted

mathematical territory. But if he succeeded, he would go down in history as the man who discovered a planet without so much as looking into a telescope. By the end of the summer holiday in October 1843, after weeks of calculation, he had arrived at a preliminary conclusion: The motion of Uranus was indeed compatible with perturbation by a large planet orbiting at twice the distance from the Sun. Adams still had to calculate the precise characteristics of this planet and its position in the sky. But he was more convinced than ever that it had to exist.

On his return to Cambridge in the autumn, Adams had to set his astronomical work to one side in order to concentrate on his college teaching duties. He was further distracted by the appearance of a comet in November 1843. The comet was observed from the Cambridge Observatory by Professor James Challis, who had succeeded Airy as the director of the Cambridge Observatory. Challis knew Adams and was aware of his particular interest in orbital calculations. So Challis asked him to calculate an orbit for the comet, based on his observations.

Adams found that the comet's orbit had taken it very close to Jupiter, and suggested that during this encounter the comet "must have suffered very great perturbations, which may have materially changed the nature of its orbit." His results were published in the *Monthly Notices of the Royal Astronomical Society* in January 1844 and were subsequently found to be in close agreement with similar calculations carried out by the French astronomer Urbain Jean-Joseph Le Verrier. This was the first of several occasions on which the paths of Adams and Le Verrier were to cross.

The following month Adams asked Challis to do him a favor in return. To pursue his investigation of Uranus, he explained, he needed as much observational data as possible. The obvious place to get it was the Royal Greenwich Observatory, given its reputation for accuracy. Rather than write to the astronomer royal out of the blue, Adams asked Challis to write a letter to Airy on his behalf, to ask for the data. Challis, who had been Senior Wrangler himself in 1825, was eager to help his promising young protégé and wrote at once.

"A young friend of mine, Mr. Adams, of St. John's College," he wrote to Airy, "is working at the theory of Uranus." Challis explained that Adams needed more data to develop his theory and was interested in observations of Uranus for the years 1818–26 in particular. The ever-efficient Airy replied at once, supplying all of the available observations from 1754 to 1830. Challis thanked him: "I am exceedingly obliged by your sending so complete a series of tabular errors of Uranus. Mr. Adams had not dared to ask for more than the errors of ten years. The list you have sent will give him the means of carrying on in the most effective manner the inquiry in which he is engaged."

Adams now had enough data to carry out a more detailed analysis of the problem. Once again he spent an entire summer engrossed in his calculations. By the autumn of 1844 Adams had, however, worked out his plan of attack.

The discrepancies between the observed and calculated positions of Uranus, he speculated, had two distinct causes: the perturbing influence of the unseen planet, and the fact that Bouvard's true ellipse for Uranus (the ellipse it would describe around the Sun if none of the other planets existed) was incorrect. So Adams derived an equation that expressed the way the characteristics of the unseen planet, and the required corrections to the true ellipse of Uranus, were related to the discrepancies in the position of Uranus. This equation was, however, so complicated that it would only just fit on a single page.

Adams was now working on the basis of two assumptions: that the unseen planet was in an elliptical (rather than a circular) orbit, and that its average distance from the sun was twice that of Uranus. He was unhappy with the second assumption, which was based on Bode's law, but he had found that unless he guessed at a value for the average radius of the orbit, the entire problem became mathematically intractable. This is because over a short period the perturbatory effect of a more distant, more massive planet would be hard to distinguish from a nearer, less massive one. Rather than deal with an infinite number of possible planets, Adams chose to make an assumption about the planet's distance first and come back and adjust it later if necessary.

Before he could proceed any farther, however, his attention was diverted by a new comet. Challis had been following it since its discovery and sent Adams his observations so that he could establish its orbit. This time, Challis was particularly keen on determining the comet's orbit before the French did; since the Astronomical Society in London (which had become the Royal Astronomical Society in 1830) was not meeting at the time, the only way to publicize Adams's results was to send them to the *Times*. Adams duly did as he was told, and his results were published in the newspaper on October 15. But he soon found out that he had been narrowly beaten by Le Verrier, who had already published his own very similar results in France.

The demands of Adams's college duties at Cambridge meant it was not until the summer of 1845 that he was able to devote himself once again to the search for his unseen planet. In June

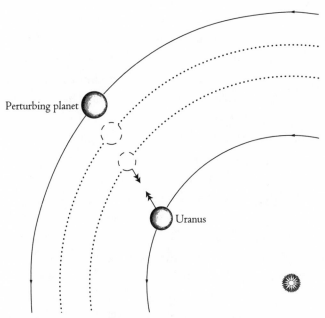

The gravitational force due to a perturbing planet increases in proportion to the planet's mass, and decreases in inverse proportion to the square of its distance. So over a short period a large, distant planet could have the same perturbatory effect as a smaller, nearer planet.

he went to the annual meeting of the British Association, a major scientific conference, where (as he wrote to his brother George) he "had the pleasure of seeing for the first time some of our greatest scientific men," including Sir John Herschel and George Airy. Just being in the same room as these great astronomers (he was too shy to actually talk to any of them) seems to have inspired Adams to mount a sustained attack on the problem, for a few weeks later he reported that he had made good progress "in calculating the place of the supposed New Planet."

The process of calculation was extremely laborious. Sitting in his rooms in St. John's College, surrounded by piles of paper, Adams started with his equation relating the discrepancy in the position of Uranus to the unknown characteristics of the unseen planet and the required adjustments to the true ellipse of Uranus. Then, by comparing the observational data supplied by Airy with Bouvard's tables, he determined the exact value of the discrepancy in 1780, 1783, 1786, and so on for every third year until 1840. Next, he wrote his equation out twenty-one times, once for each of these years, substituting the appropriate values for the discrepancy and the year into each one.

At this point, the solution of the problem was literally in Adams's hands. One equation on its own was insufficient to determine the characteristics of the unseen planet; but each different equation subtly constrained the range of possible answers. Taken together, the twenty-one equations narrowed down the solution so that, collectively, they encapsulated the answer to the strange behavior of Uranus. It was just a matter of manipulating them in the right way to make them give up their secret.

To further complicate matters, however, the equations were

$$
\begin{aligned}
c'' = {} & \partial e + \partial x_1 \cos\{13°0.5'\}t + \partial x_2 \cos\{26°1.0'\}t \\
& t\partial n + \partial y_1 \sin\{13°0.5'\}t + \partial y_2 \sin\{26°1.0'\}t \\
& + h_1 \cos\{8°24.6'\}t + h_2 \cos\{16°49.2'\}t + h_3 \cos\{25°13.8'\}t \\
& + k_1 \sin\{8°24.6'\}t + k_2 \sin\{16°49.2'\}t + k_3 \sin\{25°13.8'\}t \\
& + p_1 \cos\{4°36.0'\}t + p_2 \cos\{3°48.6'\}t + p_3 \cos\{12°13.2'\}t \\
& + q_1 \sin\{4°36.0'\}t + q_2 \sin\{3°48.6'\}t + q_3 \sin\{12°13.2'\}t
\end{aligned}
$$

Adams's equation. The discrepancy (c'') between the observed and calculated positions of Uranus is related to eighteen unknown constants (∂e, ∂x_1, ∂x_2, ∂_n, ∂y_1, ∂y_2, h_1, h_2, h_3, k_1, k_2, k_3, p_1, p_2, p_3, q_1, q_2, q_3) for twenty-one values of t, each of which corresponds to a particular year. The eighteen unknown constants are themselves combinations of the unknowns giving the corrections to the true ellipse for Uranus and the characteristics of the unseen planet.

known to be slightly inaccurate, since they were based on astronomical observations, which can never be entirely error-free. Fortunately for Adams, finding the best solution within a large number of slightly inaccurate constraints was one of the things Carl Friedrich Gauss had shown how to do in his book *Theoria Motus*. So, using Gauss's mathematical tools, Adams fought his way through this mathematical thicket, and, one day in late September 1845, he finally arrived at a solution: a set of numbers that represented the corrections to the orbit of Uranus, and the orbital characteristics of the unseen planet.

But would this solution really explain the anomalous motion of Uranus by eliminating the discrepancies between its computed and observed positions? His pen scratching across the page, Adams adjusted the orbit of Uranus using the corrections given by his solution, and then recalculated the discrepancies, this time taking into account the gravitational influence of the unseen planet. Just as he had hoped, the discrepancies vanished, shrinking from an unacceptable 90 arcseconds to a mere 1 or 2 arcseconds. As a double-check, Adams extended his predictions of Uranus's position back in time, to see whether his solution was supported by the pre-1781 observations that had been discarded by Bouvard. To his delight, he found that it was.

Adams had achieved his goal: He had shown that the errant motion of Uranus could indeed be explained by assuming the existence of an unseen, as-yet-undiscovered planet. In a sense, the planet was no longer undiscovered; Adams had found it among his pages of calculations. And now that he knew characteristics of the planet's orbit, he could determine its position in the sky.

The planets move through the constellations of the zodiac,

never straying far from a line along the middle of the zodiac called the ecliptic. Therefore, to specify the position of a planet in the sky, it is only really necessary to give its position along the ecliptic—its longitude. Adams calculated the unseen planet's longitude relative to the Sun on October 1, 1845, to be 326.5°, placing it in the constellation of Aquarius, near the border with the neighboring constellation of Capricornus.

Having established where the planet was lurking, Adams decided the time had come to visit Airy in person, to ask him to organize a telescopic search for it. He turned to Challis, who wrote him a letter of introduction to Airy, dated September 22:

Star map showing Adams's prediction of the position of the unseen planet. The line is the ecliptic, the path followed by the planets through the zodiac.

"My friend Mr. Adams (who will probably deliver this note to you) has completed his calculations regarding the perturbation of the orbit of Uranus by a supposed ulterior planet, and has arrived at results which he would be glad to communicate to you personally, if you could spare him a few moments of your valuable time. His calculations are founded on the observations you were so good to furnish him with some time ago; and from his character as a mathematician, and his practice in calculation, I should consider these deductions from his premises to be made in a trustworthy manner. If he should not have the good fortune to see you at Greenwich, he hopes to be allowed to write to you on this subject."

Despite his fulsome recommendation of Adams, Challis declined to look for the planet himself, even though Adams had told him its position. The powerful Northumberland telescope would have had no difficulty resolving the disk of a planet with the characteristics calculated by Adams; he had calculated its mass to be about three times that of Uranus, so even though it was twice as far away, the new planet would appear at least half as big, and its disk ought to be clearly distinguishable from a star. Adams even gave Challis a copy of his results, including the planet's position and an estimate of its brightness. Adams's implication was clear, but he stopped short of explicitly asking Challis to look for the planet. He later wrote that he "could not expect that practical astronomers, who were already fully occupied with important labours, would feel as much confidence in the results of my investigation as I myself did."

Adams was absolutely right. Challis subsequently explained that he was reluctant to look for the planet because it was "so novel a thing to undertake observations in reliance upon merely theoretical deductions, and that while much labour was certain, success appeared doubtful." Indeed, no astronomer in history had ever been asked to do what Adams was suggesting: to look in a particular location with the expectation of seeing an object whose existence had been deduced solely on the basis of gravitational theory. So Challis decided to leave to Airy the question of what action should be taken.

Adams arrived in Greenwich at the end of September without having made an appointment to see Airy, and he was disappointed to discover that the astronomer royal was away in France, attending a meeting of the Academy of Sciences. He left the letter from Challis and continued on his journey to Cornwall, where he spent a few weeks on holiday. Airy read the letter on his return and wrote to Challis, asking him to "mention to Mr Adams that I am very much interested with the subject of his investigations, and that I should be delighted to hear of them by letter from him."

Anxious to present his case in person, Adams decided to pay another visit to Greenwich, rather than write a letter. Before setting out, he wrote a brief summary of his results. "According to my calculations, the observed irregularities in the motion of Uranus may be accounted for by supposing the existence of an exterior planet, the mass and orbit of which are as follows," he noted, and then appended the details of the new planet. He also included a table showing how the errors in the calculated position of Uranus all but vanished when the new planet was taken into account.

On October 21, 1845, a crisp, clear autumn day, Adams walked through Greenwich Park and up the hill to the observatory, arriving at about three o'clock in the afternoon. Once

again, he had not arranged his visit in advance, and upon inquir-
ing after Airy, was told that he was out. Adams said he would
call back a little later, and left his visiting card, together with the
summary of his results, written on a single sheet of paper and
folded into eighths—an admittedly unlikely harbinger of a new
planet. After walking around for an hour or so, Adams returned
to the observatory at about four. For some reason Airy had not
been told of Adams's intention to return, and since the Airy
household dined at the unusual hour of half past three on the
order of Airy's doctor, Adams was told by the butler that Airy
was dining and could not be disturbed.

A modest and retiring fellow, Adams was hardly the sort of
person to insist that he should be admitted; he may even have
felt slighted, since he had made clear his intention to call back
later, and there was not so much as a note of acknowledgment
waiting for him from Airy. Adams turned on his heel and went
back to Cambridge. Having placed his results in the hands of the
most senior astronomer in the country, Adams hoped that Airy
would swiftly arrange for a sufficiently powerful telescope to be
directed toward the right part of the sky in order to find the
planet.

He hoped in vain. Just as he had told his brother Thomas,
the harvest does not immediately follow the sowing.

The Master Mathematician

Astronomy is pre-eminently the science of order.

—GEORGE AIRY

For someone as obsessed with order and regularity as George Airy, October 1845 was a particularly trying month. On his return from France at the end of September, he was greeted with a mountain of correspondence. His wife, Richarda, was about to give birth to their ninth child, a cause of much concern to Airy, since her previous deliveries had been difficult. He was also heavily burdened with his work for the Railway Gauge Commission.

But most worrying of all were the awful rumors suggesting that one of his employees had committed a terrible crime.

William Richardson, a junior assistant at the observatory for twenty-eight years, was said to have fathered a child by his own daughter, murdered the infant by poisoning it with arsenic, and then buried the body in his garden. On October 27 Airy's journal records that he "investigated a very serious charge of incest against Mr. Richardson, and suspended him from his office." Richardson was immediately taken into custody by the police.

When the newspapers got hold of the story, they missed no opportunity to point out the great disparity between Richardson's respectable former position as an employee of the observatory and the terrible nature of the crimes of which he stood accused. Even the *Times* repeatedly referred to Richardson as "late assistant to Professor Airy, of the Royal Observatory," implying that Airy had failed to notice a murderer right under his nose. This was unfair, because it was Airy himself who had dismissed Richardson and brought the case to the attention of the authorities. Even so, the anticipation of a trial lasting several weeks, with sensational reports in the press and the good name of the Royal Observatory being repeatedly sullied, was a mortifying prospect for the astronomer royal, who had put in years of hard work to restore the observatory's reputation. (In fact, Richardson was eventually acquitted of murder, on the grounds of inconclusive evidence, but not until the following May.)

Even amid all of these goings-on, Airy still found time to reply to John Couch Adams's note, though not until two weeks after it had been delivered. His reply, dated November 5, had a cautious tone; he wanted clarification of a particular point before taking Adams's prediction seriously.

Airy was skeptical for a number of reasons. First, he had previously expressed his belief that the irregular motion of Uranus
could be explained without recourse to an unknown planet by a
further analysis of the perturbations caused by Saturn. And even
if an unknown planet was to blame, Airy had suggested that the
calculation that Adams had performed—namely, working out its
position—was impossible. So the very existence of Adams's prediction was a direct affront to Airy's stated opinions.

Furthermore, since Adams had not had the chance to present
his work to Airy in person and had only presented his final results
rather than his entire analysis, Airy was unaware of the detailed
nature of Adams's calculations. For all Airy knew, Adams might
have merely guessed the characteristics of the new planet, rather
than calculating them from the discrepancies in the position of
Uranus. And even if he had calculated the characteristics of the
new planet from scratch, there was, thought Airy, the possibility
that an arithmetical error along the way would have resulted in
an incorrect answer.

But Airy realized that there was a way to double-check the
validity of Adams's calculations. Alexis Bouvard's flawed tables
predicted both the longitude and the radius vector (the distance
of the planet from the Sun) for Uranus, and while most people
were concerned with the errors in the predicted longitude, Airy
had gone to particular trouble to show that Bouvard's predictions
of the radius vector were wrong too. So the question was: Could
Adams's new planet also explain the error in the radius vector of
Uranus? If it could, there would be much more reason to take
his prediction seriously.

"I am very much obliged by the paper of results which you
left a few days since, shewing the perturbations on the place of

Uranus produced by a planet with certain assumed elements," Airy wrote to Adams. "But I should be very glad to know whether this assumed perturbation will explain the error of the radius vector of Uranus. This error is now very considerable."

To Adams, however, this question seemed trivial. There was, he believed, a very simple explanation for the errors in the radius vector: Bouvard's orbit was wrong because it failed to take into account the effect of the unseen planet. If the orbit was wrong, then of course its predictions of the longitude and radius vector would be wrong too. So although the question of the radius vector seemed crucial to anyone who (like Airy) doubted the existence of an unseen planet, it seemed trivial to anyone who (like Adams) assumed that the planet had to exist.

In any case, by the time Airy's letter arrived, Adams had embarked on a second calculation to improve the accuracy of his prediction. He was still unhappy about having based his calculations on the assumption, derived from Bode's law, that the new planet was twice as far from the Sun as Uranus. Would a different value reduce the errors still further and provide a more accurate prediction of the new planet's position? To answer this question, Adams resolved to redo his calculations with a slightly smaller value for the new planet's orbital radius. He decided to defer writing to Airy again until he had completed this calculation.

Airy was surprised that he didn't receive an immediate reply to his letter and may have concluded that his query had revealed a fatal flaw in Adams's calculations. At any rate, no telescope was directed toward the part of the sky where the new planet was supposedly located. Instead, Adams's prediction gathered dust in Airy's filing system.

Meanwhile, astronomers in France were showing renewed interest in Uranus. In the autumn of 1845, just as Adams was completing his initial calculation, Eugène Bouvard presented his new tables of the position of Uranus to the Academy of Sciences in Paris. The tables were the product of a decade's work done at the behest of his uncle Alexis Bouvard, who had drawn up the infamous tables of the planet's position two decades earlier, and who had died in 1843. Unlike his uncle's original tables, which had controversially discarded the pre-1781 observations of Uranus, Eugène's new tables took into account a handful of prediscovery observations. But they were still, even by his own admission, horribly inadequate, since they suggested that some of the modern observations were off by as much as an implausible 15 arcseconds. François Arago, director of the Paris Observatory and Airy's counterpart in France, decided that something had to be done. He came to the conclusion that one man in particular was the ideal candidate to sort out the troublesome planet: Urbain Jean-Joseph Le Verrier.

Le Verrier was the fast-rising star of French analytical astronomy. Born at Saint-Lô in Normandy in 1811, he had a gift for mathematics and won a place at the prestigious Ecole Polytechnique in Paris. After graduation he started his scientific career as a researcher in chemistry, a field in which he soon became an expert. But he continued to dabble in mathematics and mathematical astronomy in his spare time. In 1837, when two teaching posts became available at the Ecole Polytechnique, one in chemistry and one in astronomy, Le Verrier was somewhat surprised when, as a result of the intervention of his patron, the chemist Joseph-Louis Gay-Lussac, he was offered the astronomical post. (Gay-Lussac thought another candidate would be better suited

Urbain Jean-Joseph Le Verrier (The *Illustrated London News*)

to the chemistry post, and was confident that Le Verrier would do just as well as an astronomer.) To be offered a post at the Ecole Polytechnique was a great honor, and Le Verrier had no doubts about his response. As he wrote to his father: "I must not merely accept, but must actively seek out opportunities to improve my knowledge. I have already begun to mount the ladder—why should I not continue to climb?" An ambitious young man, Le Verrier abandoned chemisty and became an astronomer practically overnight.

Installed in his new post, Le Verrier looked around for a

problem with which to establish his credentials. He had published a paper on shooting stars in 1832, but now he needed a
more heavyweight subject. Before long, he found one: His first
foray into celestial mechanics was a masterful investigation of the
long-term stability of the solar system. During the eighteenth
century, Pierre-Simon Laplace had shown that the orbits of the
planets varied slightly as a result of the planets' mutual gravitational perturbations, but that the solar system as a whole was
stable: The planets were in no danger of being flung into space,
or into the Sun, as a result of these small oscillations. Le Verrier
decided to examine the question of exactly how much each orbit
varied, and how quickly.

In 1840 he published a paper in which he detailed the slow
tilt and drift of the orbits of the inner planets (Mercury, Venus,
Earth, and Mars) at 20,000-year intervals from 100,000 B.C.E. to
C.E. 100,000. He followed this up with another paper in which
he performed a similar analysis for the outer planets (Jupiter,
Saturn, and Uranus). And in 1844 he presented a paper to the
Academy of Sciences on the subject of cometary orbits, in which
he examined Jupiter's ability to influence the orbits of comets
and even to swallow them up altogether. It was this paper that
first brought Le Verrier to the attention of Adams, who had performed similar calculations of his own at around the same time.

Le Verrier was unquestionably a gifted theoretical astronomer. But in other respects, he was very different from Adams. In
the autumn of 1845 Adams was an unknown twenty-six-year-old
graduate student who had never published a scientific paper, and
who was working on the problem of Uranus in his spare time;
Le Verrier, eight years his senior, already had an established reputation in scientific circles, a full-time post as an astronomer, and

several papers to his credit. Furthermore, whereas Adams was modest and retiring, Le Verrier was confident and ambitious. Indeed, Le Verrier was regarded by some of his contemporaries as rather too confident and ambitious for his own good, so that he made enemies more easily than friends. His character was generally described as "difficult," and he was regarded as a *mauvais coucheur* (an awkward customer or, literally, a bad bedfellow). One contemporary said of him: "I do not know whether M. Le Verrier is actually the most detestable man in France, but I am quite certain that he is the most detested."

Detested or not, by examining the ways in which different bodies in the solar systems could influence each other's orbits, Le Verrier had shown himself to be an expert in the field of gravitational perturbations. It was for this reason that Arago suggested, in the late summer of 1845, that he turn his attention to the problem of Uranus. (The French name for the planet, Herschel, had largely fallen into disuse by this time.) Le Verrier knew that to be set such a challenge by the most senior astronomer in France was an opportunity for him to distinguish himself. He got to work immediately and presented his first paper on the subject of Uranus to the Academy of Sciences on November 10, 1845.

By this time Adams's solution to the problem was already in Airy's hands. But only Adams, Airy, Challis, and a handful of people in Cambridge knew about it. Unaware that his English rival had already got as far as predicting the position of a perturbing planet, Le Verrier approached the problem one step at a time. His first paper dealt simply with the question of whether the motion of Uranus could be explained by a more detailed analysis of the perturbations caused by Jupiter and Saturn. Using two separate mathematical arguments, Le Verrier examined the influence of Saturn in particular, and in greater detail than anyone (including Adams) ever had before. In fact, he calculated Saturn's influence on the position of Uranus to within $1/20$ of an arcsecond, or $1/72,000$ of a degree. In the process, he uncovered a number of rather embarrassing inconsistencies in Alexis Bouvard's tables, thus discrediting them even further. But his conclusion was unequivocal: Perturbations alone were insufficient to explain the anomalous motion of Uranus.

Le Verrier's paper reached Britain in December, and Airy was extremely impressed by it. He had previously expressed the belief that it might be possible to explain the motion of Uranus by a more thorough analysis of the perturbations; now he had to admit that this was not the case. But Le Verrier's paper did nothing to induce Airy to take Adams's results more seriously. For his part, Adams was by this time hard at work on the refined version of his prediction, unaware that Le Verrier was now working on the same problem.

On June 1, 1846, Le Verrier published his second paper on the problem of Uranus. The paper was divided into two parts. If the treatment of the perturbations was not responsible for the incorrect position of the planet, he asked in the first part, could the errors in its position be remedied by adjusting the "true" ellipse of its orbit? Le Verrier took 19 "ancient" observations and 260 modern ones made at the Greenwich and Paris observatories, re-reduced them all, and then recalculated the errors. Next, he devised an equation relating the error for each observation to the corrections to the true ellipse—just as Adams had done, only

Adams had also included terms for a perturbing planet too. But Le Verrier was not interested in a new planet just yet; he wanted to see if adjusting the ellipse alone could explain the errors. Boiling down his 279 equations, he calculated the adjustments to the ellipse that would minimize the errors. He then adjusted the ellipse and recalculated the errors. They were still significant. Le Verrier had thus shown beyond doubt that something other than an incorrect orbit was responsible for the errors in the predicted position of Uranus; a new hypothesis was required.

In the second part of the paper, Le Verrier considered other possible causes for the planet's mysterious motion. He began by declaring that the idea that Newton's law of gravitation might not apply at Uranus's great distance from the Sun was only to be considered as "a last resort to which it cannot be permitted to have recourse without having exhausted the examination of other causes, and having shown them to be unable to produce the observed effects." Dismissing the cometary impact, resistive medium, and massive satellite theories, he concluded that "no other hypothesis remains except that of a perturbing body continually acting upon Uranus, changing its motion in a very gentle fashion. Given what we know about the constitution of the solar system, this body can only be a planet, hitherto unknown. But is this hypothesis any more credible than the others? Is anything about it incompatible with the observed errors? Is it possible to work out the position that this planet would occupy in the sky?"

Like Adams, Le Verrier decided to work on the assumption that the planet must orbit the sun at twice the distance of Uranus, invoking the "singular law which has revealed itself in the average distances of the planets from the sun." The planet had to be beyond Uranus, he argued, otherwise it would also affect the orbit of Saturn; if it were any more distant—at, say, three times the distance of Uranus from the Sun—it would have to be so massive in order to influence Uranus from so far away that its effects on Saturn would also be noticeable. Le Verrier thus followed Bode's Law, though he avoided actually mentioning its name.

Le Verrier then reformulated his equations to take into account the influence of the perturbing planet. His approach was, however, different from that of Adams. Le Verrier manipulated his equations to produce a formula relating the mass of the unknown planet to its position in the zodiac (i.e., its longitude) in the year 1800. He then considered forty positions in the sky where the planet might have been in 1800, equally spaced around the zodiac at 9° intervals. For each one, he worked out what the mass of the planet would have to have been in order for it to have perturbed Uranus in the manner observed.

Clearly, any position in the sky for which the formula yielded a negative mass could be eliminated, because a negative mass is impossible. Similarly, any position for which the mass was found to be very large could also be eliminated, because a very massive planet would have had a noticeable effect on Saturn. In this way, Le Verrier was about to narrow down the area of the zodiac containing the planet, concluding that "there is only one region . . . in which it is possible to place the disturbing planet so as to account for the movements of Uranus. . . . the mean longitude of this planet, on January 1, 1800, must have been between 243° and 252°."

Next, Le Verrier performed a more detailed analysis of five candidate positions of the planet around this region, to see which one minimized the errors in the position of Uranus, and he found that 252° gave the best result. He then roughly worked out the planet's orbit and calculated where it would be on January 1, 1847: at a longitude of 325°, on the border between Aquarius and Capricornus. This result was, admittedly, preliminary, but Le Verrier was confident that it had to be accurate to within 10° at the very worst. "Such is the capital result at which I have arrived," he concluded with characteristic self-confidence. There was, he said, no longer any doubt that the planet had to exist, since he had shown that its existence "perfectly accounts for the observed errors of Uranus."

All that remained, added Le Verrier, was to calculate the new planet's orbit more precisely—something he planned to do in his

Star map showing Le Verrier's prediction of the position of the unseen planet.

next paper on the subject. No doubt, he warned, some people would want to limit the solar system to its current size, rather than admit the existence of a new planet. "In which case, however, I would reply that one might have had the same reason to believe, on March 12, 1781, that Saturn was the most distant planet, only to have been contradicted the following day by the discovery of Uranus."

The positions worked out by Le Verrier and Adams for the new planet corresponded extremely closely. Working independently, unaware of each other's calculations, and using totally different methods, the two men had come to the same conclusion. It was as though they had both scaled the same mathematical mountain from different sides: Adams by hacking his way unseen through the undergrowth on the south face, and Le Verrier by methodically scaling the north face, in full view of the scientific community.

Le Verrier's second paper received a rapturous welcome across Europe. Not only was it the first published account of an investigation into the possibility of a new planet, it was also comprehensive and conclusive. Airy received his copy at Greenwich in the last week of June. "I cannot sufficiently express the feeling of delight and satisfaction which I received from it," he later remarked. At this point, he alone knew that Le Verrier's prediction agreed closely with that of Adams. This gave Adams's prediction a new credibility and meant that Airy was now far more prepared to consider the idea of an unseen planet.

On June 25, Airy wrote to his friend William Whewell, a senior Cambridge academic who was updating a science book and wanted to know the latest thinking about the anomalous behavior of Uranus. "People's notions have long been turned to the effects of an external planet," reported Airy, "and upon this there are two remarkable calculations. One is by Adams of St. John's (which in manuscript reached me first). The other is by Le Verrier. Both have arrived at the same result, viz. that the present longitude of the said disturber must be somewhere near 325°."

Airy still, however, wanted to know whether the supposition of a new planet explained the errors in the radius vector of Uranus. Just as he had written to Adams eight months earlier, he now addressed the same query to Le Verrier. "I have read, with very great interest, the account of your investigations on the probable place of a planet disturbing the motions of Uranus; and I now beg leave to trouble you with the following question," he wrote to Le Verrier on June 26, in a letter that makes no mention whatsoever of Adams. Would the observed errors in the radius vector, Airy asked, "be a consequence of the disturbance produced by an exterior planet, now in the position which you have indicated? I imagine that it would not be so." Airy explained that, as far as he could see, the exterior planet would not explain the error in the radius vector. This meant that his letter was a direct challenge to Le Verrier, in a way that his letter to Adams had not been.

On June 29, Airy addressed a meeting of the Board of Visitors of the Royal Observatory. This meeting was an annual event at which a report on the observatory's work was presented to a scientific committee chaired by the president of the Royal Society. Airy told the board, whose members included his fellow Cambridge astronomers Sir John Herschel and James Challis, of the "extreme probability of now discovering a new planet in a very short time, provided the powers of one observatory could be directed to the search for it." He later wrote that he gave "as the reason upon which this probability was based, the very close coincidence between the results of Mr Adams' and Mr Le Verrier's investigations of the place of the supposed planet disturbing Uranus."

Two days later, Airy received a reply from Le Verrier. "I am just about to finish the correction of the elements of the perturbing planet," Le Verrier explained, "and to work out its position with great precision. If I might be allowed to hope that you will have enough confidence in my work to look for this planet in the sky, I will hasten, Sir, to send you its exact position, as soon as I have obtained it." Turning to the radius vector question, Le Verrier explained that of course the radius vector was wrong because Bouvard's true ellipse was wrong. Using his corrected orbit for Uranus, he said, the error disappeared entirely. "It is, in fact, one of the considerations which must increase the likelihood of the truth of my results, that they scrupulously account for all aspects of the problem. . . . The radius vector is corrected on its own, without having to be considered separately. Excuse me, Sir, for insisting on this point."

This letter quashed any last doubts Airy might have had about the existence of the new planet. There was no need to wait for Le Verrier's more accurate calculations; the fact that his first approximation agreed so closely with Adams's prediction suggested that it was near enough. Without replying to Le Verrier, Airy decided that the time had come to begin the search for the planet.

The astronomer royal's confidence that a new planet was

about to be discovered, as expressed to the Board of Visitors, made a strong impression on John Herschel. Around this time Herschel was writing a speech to be delivered at the meeting of the British Association a few weeks later. He was inspired to write the following passage about the new planet: "We see it as Columbus saw America from the shores of Spain. Its movements have been felt, trembling along the far-reaching line of our analysis, with a certainty hardly inferior to that of ocular demonstration." His audience would, of course, assume that Herschel was referring simply to Le Verrier's published calculations. But a handful of people would know that, thanks to Adams, the astronomers of England had even greater reason to believe in the existence of the new world—and a head start in the effort to identify it among the stars.

7

The Noblest Triumph of Theory

A new planet, it is said, has lately been discovered. This is not correct. The planet is as "old as the hills."

—*SCIENTIFIC AMERICAN,* 1846

Through a telescope, there are two ways to distinguish a planet from a star: by its disk, and by its motion against the starry background, noted over the course of several observations. So any astronomer with a prediction of the existence of a new planet could look for it in one of two ways. The first would be to examine the stars in the area of the sky where the planet was thought to be, and look for a "star" with a visible disk. The second approach would be to record the positions of the stars in the vicinity of the planet's predicted position, repeat the process a few days later, and see if any of the stars had moved.

When George Airy resolved to launch a search for the planet, he settled upon this second approach, since it was more thorough and did not rely on any assumptions about the planet's size. Because no sufficiently detailed star map of the area was available, Airy's plan entailed drawing up an entirely new map of the stars in the vicinity of the planet's suspected position and then checking the position of each star a few weeks later. Airy reckoned that a third sweep over the area in question would also be necessary, to provide a double-check of the positions recorded during the first two sweeps.

Having decided on this plan, however, Airy did not rush to the nearest telescope and start searching for the planet. Nor did he instruct any of his staff at Greenwich to do so. Instead, he concluded that the most suitable telescope with which to conduct the search was the Northumberland telescope at the Cambridge Observatory, which he had installed during his tenure as professor of astronomy, and which was now under the charge of James Challis. Airy's decision to hand responsibility for the search back to Challis in Cambridge seems to have been based on a number of considerations.

First, the Northumberland, with its 11.75-inch aperture, was far larger and more powerful than any telescope at the Royal Greenwich Observatory. For the day-to-day duties of measuring the positions of stars and planets at Greenwich, the accurate alignment of the telescopes was more important than their magnifying power, so the largest instrument at the Royal Observatory was a refracting telescope with an aperture of 6.7 inches. Given that the brightness of the new planet was unknown, it made sense to search for the planet with the largest available telescope. (John Couch Adams had, in fact, calculated that the new planet ought to be visible even in small telescopes, but Airy was unaware of this.)

Second, conducting the search from Greenwich would have disrupted the observatory's routine to an unacceptable degree. Airy remarked in a letter to his friend William Whewell that if "I were a rich man or had an unemployed staff I would immediately take measures for the strict examination of that part of the heavens containing the position of the postulated planet." But Airy would not have been able to spare the two or three staff members needed to map the part of the sky in question without compromising the observatory's routine work. Besides, given his strong views on the purpose of the Greenwich observatory, Airy would no doubt have regarded the search as an inappropriate use of observatory resources.

It also seems likely that Airy wanted the new planet to be discovered in Cambridge in particular. Along with Adams and Challis, Airy was a graduate of Cambridge University and would have been eager to secure the telescopic discovery of the planet from the Cambridge Observatory, following the prediction of a Cambridge scholar.

Airy's desire for a Cambridge discovery would also explain why he did nothing to publicize Adams's prediction, even after its confirmation by Le Verrier. For although Airy referred to Adams's work at the meeting of the Board of Visitors (which was largely made up, in any case, of Cambridge men), he failed to mention it in his letter to Le Verrier asking about the radius vector. He also seems not to have mentioned it to Peter Han-

sen—a Danish astronomer who was a leading authority on gravitational perturbations and who had previously corresponded
with Alexis Bouvard on the subject of Uranus's strange behavior—even though Hansen stayed with Airy for several weeks
starting in June 1846. Airy and Hansen even bumped into Adams
by chance while on a day trip to Cambridge on July 2; Airy
moved Hansen along after a brief exchange of pleasantries, pre-
sumably because he did not want Adams to let the cat out of the
bag.

A search from Cambridge would also have appealed to Airy
for another reason. The Northumberland telescope was so
named because it had been paid for with funds donated by the
duke of Northumberland; Airy, who had designed the telescope
and supervised its installation, might also have wanted to return
the duke's favor by using the telescope to make what was certain
to be a historic discovery.

Having determined that Cambridge was the best place from
which to conduct the search, Airy wrote to Challis on July 9.
"You know that I attach importance to the examination of that
part of the heavens in which there is a possible shadow of a reason for suspecting the existence of a planet exterior to Uranus,"
he began. There was, he declared, "no prospect whatever of [the
search] being made with any chance of success except with the
Northumberland telescope." Airy asked Challis if he would undertake the search, and offered to supply an assistant from the
Royal Greenwich Observatory to help.

Airy's letter had an uncharacteristically hasty tone. "You will
readily perceive that all this is in a most unformed state at present, and that I am asking these questions . . . in the hope of
rescuing the matter from a state which is, without the assistance

that you and your instrument can give, almost desperate," he wrote. Four days later, having received no response from Challis, Airy wrote again, giving details of how the search should be carried out. The tone of this letter was also at odds with Airy's usual calm, unhurried demeanor. "I have drawn up the enclosed paper in order to give you a notion of the extent of work incidental to a sweep for the possible planet," he wrote. "I only add at present that in my opinion, the importance of this inquiry exceeds that of any current work, which is of such a nature as to be totally lost by delay."

Why was Airy in such a hurry for the search to begin? His plan for the search proposed using the Northumberland in a

Star map showing the extent of the search area proposed by Airy.

similar manner to a transit telescope: It was to be pointed at
the appropriate part of the sky and fixed, so that the rotation of
the Earth would carry the stars in the search area across the field
of view, and the position of each star could be precisely noted.
Although Airy was confident that Adams and Le Verrier had collectively shown that a new planet had to exist, he was less confident in their ability to predict its exact position. So he proposed
an enormous search area, 30° long and 10° wide, centered on
the predicted position and extending across the constellations
Aquarius and Capricornus. To cover this area with a transit
telescope would, Airy estimated, require eighty separate "sweeps,"
each covering a small strip of the search area, and each taking
more than an hour, with the telescope's position slightly adjusted each time. As each star crossed the telescope's crosshair,
its position would be called out by the observer and recorded,
along with the time, by an assistant. It would then be possible to
draw up a map of the search area by plotting the position of each
star. The entire process would, Airy calculated, require around
300 hours of observing time. Given the unreliability of the
British weather, the search looked likely to take several months.

But anyone else looking for the planet would be faced with
the same task, so the Cambridge Observatory would have a head
start of several weeks if it began the search before Le Verrier
published his refined prediction of the new planet's position.
Having found the planet, Airy and Challis could then announce
that the search had been inspired by Adams's prediction of the
previous year, for which Le Verrier's work had subsequently provided independent confirmation. The result would be a double
triumph for British astronomy, and for Cambridge in particular.

On July 18, after returning from a holiday, Challis replied to

Airy that he intended to start the search as soon as possible, even though he already had his hands full with the reduction of a large number of cometary observations. No matter; he could carry out sweeps for the new planet at night and work on his cometary calculations by day. Challis informed Adams, who eagerly drew up a table giving the planet's expected position between July 20 and October 8.

Whether or not Airy and Challis formally agreed to keep Adams's prediction secret is unclear, but it is remarkable that neither of them breathed a word about it to any of Britain's amateur astronomers, many of whom were equipped with telescopes just as powerful as the Northumberland, and who might quickly have located the planet. Previously, Challis had encouraged Adams to write to the *Times* to publicize his analysis of cometary orbits; but he did not make the same suggestion in this case. Instead, both he and Airy seem to have done their utmost to ensure that the discovery would be made in Cambridge, and with the Northumberland telescope.

In any event, Adams was delighted that his prediction was finally being taken seriously. On July 29, 1846, five years after he had first set himself the task of explaining the anomalous motion of Uranus, and nine months after he had delivered his results to Airy, a telescope was finally directed toward the part of the sky indicated by his calculations.

On the first night of the search, Challis concentrated on recording the positions of stars in the center of the search area, since that was where the planet was most likely to be found. He went on to make further observations on July 30 and August 4 but was prevented from observing by cloud and moonlight (which makes faint stars difficult to see) until August 12, when he went back over the area he had covered on July 30.

After this preliminary series of observations, Challis decided to check that his observing method really worked by comparing the results from July 30 and August 12 to ensure that they agreed. After comparing the positions of thirty-nine stars from the two nights, all of which agreed perfectly, he was confident that his approach would eventually flush out the planet. He then continued his observations throughout August, recording the positions of hundreds of stars in the search area. "I have lost no opportunity of searching for the Planet, the nights having been pretty good. I have taken a considerable number of observations," he wrote to Airy on September 2. "But I get over the ground very slowly, and I find that to scrutinise thoroughly in this way the preferred portion of the heavens will require many more observations than I can take this year."

Challis had, in fact, already glimpsed his quarry. Had he compared just ten more star positions from his August 12 observations with those of July 30, he would have noticed that the forty-ninth star seen on August 12 was missing from his July 30 observations. Without realising it, Challis had the planet in his grasp.

The Northumberland telescope was not the only telescope sweeping across Aquarius and Capricornus in search of a new planet during the summer of 1846. Speculation about "Le Verrier's Planet" had become a matter of general public interest. An article published in the *Times* on August 4, titled "The New Planet," quoted from the French newspaper *Le Constitutionnel* and expressed the commonly held view that it would only be a matter of time before the planet was found. This would, said the paper, be a great triumph of theory: "Analysis transports us to the regions of the unknown, and brings us back laden with the most splendid discoveries. . . . Let us hope that if chance discovered Uranus, we shall soon succeed in seeing the planet whose position has been ascertained by M. Le Verrier." Le Verrier's prediction was similarly reported, as a novelty item, in other newspapers and journals, complete with the predicted value of the planet's supposed longitude.

In France, a brief search was mounted from the Paris Observatory but was abandoned by early August when no planet was found and the extent of the star-mapping work that would be required to find it became apparent. François Arago, the director of the observatory, was, like his counterpart Airy, reluctant to devote official resources to the search. Pressure of official work also hampered a search for the planet proposed in August by Sears Cook Walker, an astronomer at the U.S. Naval Observatory in Washington, D.C., who was told that there was no room in the observational schedule until October at the earliest.

In London, a few sweeps were carried out by John Russell Hind, a former assistant at the Royal Greenwich Observatory, who was in charge of a private observatory in Regent's Park equipped with a 7-inch refractor. Hind and Challis had written to each other about the prospect of finding a new planet, and Hind was dimly aware of Adams's interest in calculating its position, but Challis had not informed him of Adams's prediction or of its agreement with Le Verrier's. So Hind's sweeps, which failed to find anything, were based solely on Le Verrier's predicted position.

Why did other astronomers, both professional and amateur, not leap to their telescopes? Many of them may simply have decided to wait. Le Verrier's prediction was rather vague, since it was, by his own admission, only accurate to within 10°. Anyone who read Le Verrier's paper would have known that he planned to embark on a further, more refined prediction of the new planet's position, and might have chosen to wait for his next paper before commencing a search.

On August 31, Le Verrier presented the Academy of Sciences with his third paper on the theory of Uranus, in which he did everything he could to encourage astronomers to search for his planet. (He was, of course, totally unaware of the search already under way in Cambridge.) Le Verrier had, by this time, spent a whole year working full-time on the problem, and his calculations covered more than 10,000 pages. His paper explained how he had derived a detailed orbit for the new planet, along with a more accurate prediction of its longitude on January 1, 1847, which he revised from 325° to 326.5°. Just as important, the paper gave his estimate for the mass of the planet as two and a half times that of Uranus, from which its size and brightness could be calculated. Le Verrier concluded that the new planet ought to appear as a disk 3 arcseconds across, or about three-quarters of the apparent size of Uranus. And this, he said, suggested the best way to search for it.

"Not only should it be possible to see the new planet in good telescopes, but also to distinguish it by the size of its disc," he observed. "This is a very important point. If the planet could be confused, by its appearance, with the stars, it would be necessary, in order to distinguish it among them, to examine all the small stars in the region of the sky to be explored and to detect the movement of one of them. This work would be long and wearisome. But if, on the contrary, the planet has a sensible disc which prevents it from being confused with a star, if a simple study of its appearance can replace the rigorous determination of the positions of all the stars, the search will proceed much more rapidly."

Just as Le Verrier was publishing his revised prediction of the planet's position in Paris, Adams was also finishing his second analysis of the problem. On September 2, Adams wrote to Airy with his new results, unaware that the astronomer royal had gone on holiday to Germany for several weeks. Adams had found that choosing a slightly smaller value for the perturbing planet's average distance from the Sun (373 units on the Bode scale, rather than 384) had the effect of reducing still further the discrepancies between the observed and calculated positions of Uranus. He gave a new prediction of the planet's longitude of 330°, based on this new hypothesis, and requested further information in order to complete a third analysis of the problem, based on an even smaller value (344 units) for the planet's average distance from the Sun. He also belatedly addressed Airy's query from the previous year about the error in the radius vector, giving detailed figures to show that his hypothesis did indeed explain it. Adams concluded his letter by informing Airy that he was drawing up a paper explaining his calculations, which he intended to make public at the annual meeting of the British Association a few days later. But when he arrived at the conference on September 15 to present his paper, he found that he was too late; the relevant part of the meeting (mathematical and physical science) had ended the previous day. So Adams's results remained unknown to the scientific community.

By this time Le Verrier was doing everything he could to

Star map showing Adams's and Le Verrier's revised predictions of the position of the unseen planet (left and right crosses, respectively) on October 1, 1846.

persuade astronomers to look for his planet. He sent a copy of his third paper to Professor Heinrich Schumacher, the editor of the German journal *Astronomische Nachrichten*, in order to give it the widest possible circulation, and explained that he was having difficulty getting any French astronomers to look for the planet. In his reply, Schumacher recommended that Le Verrier write directly to a couple of astronomers with particularly powerful telescopes. In particular, Schumacher suggested Friedrich Struve, a German astronomer based at Pulkovo, near St. Petersburg, and Lord Rosse, a British amateur astronomer. (Lord Rosse had continued the tradition of constructing large reflecting telescopes where William Herschel left off, and in 1845 he completed a giant telescope with a mirror 6 feet in diameter.)

Possibly as a result of Schumacher's suggestion, Le Verrier suddenly remembered a letter he had received the previous year from Johann Gottfried Galle, an assistant at the Berlin Observatory. Galle had sent Le Verrier a copy of his dissertation, which consisted of reductions of planetary observations made by the Danish astronomer Olaus Rømer in 1706, because he thought Le Verrier might find it useful. Realizing that the Berlin Observatory was equipped with a powerful telescope (the 9-inch Fraunhofer refractor), which might be able to pick out the disk of his predicted planet, Le Verrier belatedly replied to Galle.

His letter, dated September 18, began with an effusive and rather ingratiating expression of gratitude for a gift that had arrived almost a year earlier. Le Verrier praised the "perfect clarity" of Galle's explanations and the "complete rigour" of his results. He then cut to the chase: "I would like to find a persistent observer, who would be willing to devote some time to an examination of a part of the sky in which there may be a planet to discover." He went on to give his predictions for the position of the planet and the size of its disk.

In writing personally to Galle, Le Verrier was being slightly devious: He was going behind the back of Johann Franz Encke, the director of the Berlin Observatory. But that, of course, was the whole point. No observatory director seemed to be prepared to spend any official time looking for the new planet; Le Verrier needed the cooperation of a more enthusiastic subordinate prepared to make observations on his own initiative. To avoid seeming discourteous, Le Verrier added a postscript asking Galle to pass on his compliments to Encke, and sent the letter to Berlin.

Galle received Le Verrier's letter on September 23, which happened to be Encke's fifty-fifth birthday. Galle was flattered that
such a famous astronomer had written to him, but Le Verrier's
request put him in a rather difficult position; he had been asked
to look for the planet as a personal favor, so he could expect
little support from Encke. Indeed, Encke had previously been
unreceptive to the idea of looking for Le Verrier's planet. But
after some badgering, Encke eventually agreed that Galle could
use the Fraunhofer telescope to look for the planet that night, on
his own. As for himself, said Encke, he planned to spend the
evening at home, celebrating his birthday.

The conversation between the two men was overheard by
Heinrich d'Arrest, an astronomy student who had taken lodgings
in one of the observatory outbuildings, so that he could gain
more practical experience. D'Arrest begged to be allowed to take
part in the search, and Galle agreed.

That night, under a beautifully clear sky, Galle and d'Arrest
turned the Fraunhofer telescope toward the place indicated by
Le Verrier—a place that Challis had, by this time, already criss-
crossed several times during the course of his own observations.
But after careful examination of the stars in the region, not one
was found to have the clear disk-like appearance characteristic of
a planet.

Disheartened, the two astronomers considered the option of
comparing the area in question with a star map. If the planet
existed, it would not be shown on a star map, because it would
not have been in the same part of the sky when the map was
compiled. D'Arrest pointed out that several volumes of a highly
detailed new star atlas, compiled by the Berlin Academy, had
recently become available, although the atlas was still incomplete

and had yet to be widely distributed. He suggested checking to see whether the volume corresponding to the area indicated by Le Verrier was among those available. Galle led him to Encke's closet and opened the drawer where the observatory's star maps were kept in a disorderly pile. Among these maps were several volumes of the new atlas, and after rummaging around for a while, d'Arrest found the volume he was looking for: the most detailed map ever published of the area of the sky thought to contain the planet.

"We then went back to the dome, where there was a kind of desk, at which I placed myself with the map, while Galle, looking through the refractor, described the configurations of the stars he saw," d'Arrest later recalled. For each star in the vicinity of the position predicted by Le Verrier, d'Arrest checked that Galle's coordinates corresponded with a star on the map. Shortly before midnight, after checking a few stars in this way, Galle described a faint star in a particular position, and d'Arrest found that there was no corresponding star in the catalog. "That star," he declared, "is not on the map!"

Gripped by the sudden possibility that they had found Le Verrier's planet, Galle and d'Arrest checked and rechecked their coordinates. There was no doubt about it; the "star" in question, at a longitude of just under 327°, had not been present when the map was compiled. After rushing to fetch Encke from his birthday party, the astronomers continued to observe the mystery object until the following morning. Upon closer inspection, they thought that perhaps it did have a disk after all, and although they could not be sure, it seemed to be moving very slowly relative to the other stars nearby. All of this suggested that it was, indeed, a new planet.

The following day, September 24, the astronomers waited anxiously for the skies to darken so that they could observe the anomalous star again. Once night had fallen, the star was once more fixed in the Fraunhofer telescope's sights, and its coordinates were carefully measured. When the coordinates were compared with those from the previous night, it was clear that the object had moved, and by exactly the amount that Le Verrier predicted that his planet would move in one day. There was no doubt that a new planet had been discovered, and within one degree of the position indicated by Le Verrier.

Now that they had established its true nature, Encke and

Star map showing Adams's and Le Verrier's revised predictions of the position of the unseen planet (left and right crosses, respectively) and the position where the planet was actually found by Galle (marked with an arrow).

Galle found the planet's disk easier to discern. ("A disc," Encke dryly remarked, "is only seen when it is known to exist.") Its diameter was measured twice by both Encke and Galle, and the average of their measurements was 2.6 arcseconds, which was reasonably close to the 3.3 arcseconds predicted by Le Verrier.

The following morning Galle wrote to Le Verrier to inform him of the discovery. "Sir," he wrote, "the planet whose position you have pointed out *actually exists*." Galle gave the position of the planet on the previous two nights and confirmed that its diameter had been measured as about 3 seconds of arc, just as Le Verrier had predicted. As the discoverer of the planet, or at least the codiscoverer, Galle felt entitled to suggest a name. "Perhaps this planet is worthy of being called Janus, after the Roman god; in addition, the double face would be appropriate for its position on the frontier of the solar system," he wrote.

Encke also wrote to Le Verrier to congratulate him: "Allow me, Sir, to congratulate you most sincerely on the brilliant discovery with which you have enriched astronomy. Your name will be forever linked with the most striking proof imaginable of the validity of the law of universal gravitation." As soon as he heard the news from Encke, Schumacher also wrote to Le Verrier. "This is," he declared, "the noblest triumph of theory that I know."

8

Possession of a New World

Le Verrier first his learned eyes upraised,
And on the problem with fixed purpose gazed;
No inward fears subdued his generous soul;
No dread of censure could his mind control;
The fame of others his bold spirit fired,
And with the hope to emulate inspired.
He passed the barriers of those distant bounds,
Once thought to mark the planets' lonely rounds,
Chasing each one in its varying course,
To each assigned its attractive force;
Planting the flag of Science wide unfurled,
Upon the flaming ramparts of the world;

And traversing the sphere by mental toil,
Returns victorious with his well-earned spoil.

—FROM A POEM PUBLISHED IN THE *LIVERPOOL*
MERCURY, DECEMBER 11, 1846

Urbain Jean-Joseph Le Verrier was thrilled to receive the news of Johann Gottfried Galle's discovery. "I thank you cordially for the alacrity with which you applied my instructions. We are, thanks to you, definitely in possession of a new world," he wrote to Galle on October 1. Even so, Le Verrier was not at all impressed by Galle's suggestion of a name for this new member of the solar system, and said so in no uncertain terms. The suggestion of the name Janus was, Le Verrier informed Galle, inappropriate because it "would imply that this is the last planet of the solar system, which we have no reason at all to believe." Besides, added Le Verrier, the decision over the new planet's name had already been taken. "The Bureau of Longitude," he informed Galle, "has decided on Neptune." The jostling over the new planet had begun.

The Bureau of Longitude had, in fact, made no such decision; it later issued a formal denial that it had played any part in deciding the name. As the publisher of the official French planetary tables, the bureau clearly had an interest in the new planet, but had no particular right to decide on a name and had never named a planet before. Perhaps Le Verrier discussed the name of the new planet on an informal basis with one or more members of the bureau, but no formal approval was issued. Instead, it appears that Neptune was the name chosen by Le Verrier himself.

To ensure that his choice prevailed, Le Verrier wrote to a number of Europe's most senior astronomers, including Airy and Gauss, to announce the discovery of the planet and give its coordinates and its name, along with the bogus endorsement of the Bureau of Longitude. Neptune was generally regarded by astronomers as an excellent choice, since it maintained the mythological tradition. That said, Neptune was a slightly odd name for the French to have chosen, since it had previously been suggested (following the discovery of Uranus) as a means of honoring Britain's naval supremacy.

As news of the discovery spread across Europe, Le Verrier was showered with praise and achieved instant fame as the man who had found a planet by calculating at his desk, rather than looking through a telescope. The normally sedate weekly meeting of the Academy of Sciences on October 5 was transformed into a public spectacle when it emerged that Le Verrier would be present, and hundreds of people flocked to the academy to see the great man in person.

The Paris newspaper *Le National* reported that "the door of the Academy, to which access is ordinarily quite easy, was blocked by crowds, and a great commotion filled the hall." The chattering of excited onlookers completely drowned out the voice of the secretary of the academy as he read the minutes of the previous week's meeting, to which nobody paid any attention at all. According to *Le National*, "Le Verrier's name was on every tongue. Where was he sitting? 'Point him out to us', visitors demanded, and, in response to a gesture from one of the Academy's old regulars, their eyes would turn towards the pale young man at the end of the green table whose health, not long ago so flourishing, had weakened under the strain of overwhelming toil. . . . The desk was covered in letters from the observatories of Europe's most illustrious astronomers, congratulating M. Le Verrier and asking him what name he wished to bestow on his planet. 'Your discovery is the most brilliant in the entire history of astronomy'—such was the gist of the correspondence."

Yet by this time, Le Verrier had changed his mind. The planet was being widely referred to, by those who had not been informed of the name Neptune, as "Le Verrier's planet." Overwhelmed by his sudden celebrity—he was, after all, the toast of

Le Verrier explains his discovery to the French king, Louis Philippe. (Mary Evans Picture Library, London)

Paris—and mindful of the fact that in France, Uranus had been known as Herschel for many years, Le Verrier decided that he did not want to call the new planet Neptune, after all. Instead, he wanted to name the planet after himself.

As arrogant and self-important as he was, however, even Le Verrier realized that to say so would be seen as breathtakingly presumptuous. So he made his feelings known to François Arago and asked him to decide on the new planet's name, while making it perfectly clear which name he wanted Arago to choose. And so, after reading aloud Galle's triumphant letter describing the discovery, Arago announced to the academy that Le Verrier had asked him to name the planet. And, said Arago, he had chosen the name Le Verrier.

This made perfect sense, Arago explained to the gathered throng, because comets were routinely named after their discoverers—so why not planets? Surely, he argued, this was the best way to honor "the name of the man who, by an admirable and unprecedented method, has demonstrated the existence of a new planet." He added that asteroids should henceforth be known by the names of their discoverers, and that Uranus should once again be known in France as Herschel.

Arago's argument was nonsense, as he must have been all too aware. It would have meant, for example, that the asteroids Juno and Vesta, both of which had been discovered by Heinrich Wilhelm Olbers, would end up with the same name. And it was the naming of Vesta, which Olbers had delegated to Gauss, that had originally established the precedent that an astronomer could ask someone else to name one of his discoveries. Even so, said Arago, "I pledge myself never to call the new planet by any other name than Le Verrier's Planet. I believe I will thus give an irrefutable

proof of my love of science, and follow the inspiration of a legiti-
mate patriotism."

The following day, Le Verrier and Arago sent off letters announcing the new name. Le Verrier also made a last-minute
change to the first page of his collected papers on Uranus, which
were then being printed, in order to fall into line with Arago's
ludicrous new planetary naming scheme. He changed the title
from *Research into the Movements of the Planet Uranus* to *Research
into the Movements of the Planet Herschel* and added a footnote
apologizing for the fact that the planet was referred to as Uranus
throughout. "In my future publications," he explained, "I shall
consider it my strict duty to avoid the name Uranus altogether,
and only to refer to it by the name Herschel. I deeply regret that
the printing of this work is already so far advanced that I am
unable to adhere to a vow that I shall observe religiously in future." The tussle over the name of the planet was, however, to
prove the least of Le Verrier's worries.

On October 1, the London *Times* printed a letter from the amateur astronomer John Hind, who explained that he had been
informed of the discovery of "Le Verrier's planet" by a friend at
the Berlin Observatory. Hind gave the coordinates of the planet
and noted that "this discovery may be justly considered one of
the greatest triumphs of theoretical astronomy." He added that
he had observed the planet himself from London on September
30, despite the strong moonlight, and that he had been able to
discern the disk. His letter enabled astronomers through Britain,
both professional and amateur, to track the planet down for
themselves.

At this point Adams's work and Challis's search were still unknown to all but a select handful of people in Cambridge. Airy was still overseas on holiday; he heard of the discovery of the new planet on September 29, while staying with Peter Hansen in Gotha. (During his trip, Airy also visited Carl Friedrich Gauss and the elderly Caroline Herschel, who had returned to Hanover after William Herschel's death in 1822.)

With the astronomer royal abroad and unavailable to make an official statement about the Cambridge search, Sir John Herschel decided it would be a good idea to put the existence of Adams's prediction into the public record as soon as possible. He also wanted to point out that he had alluded to the imminent discovery in his speech to the British Association a few weeks before. On October 3 he wrote to a weekly London newspaper called the *Athenaeum*, naming Adams and making his role in the matter public for the first time.

Only now, wrote Herschel, could he explain the whole story behind his remark, made in his speech the previous month, that the new planet could be seen "as Columbus saw America from the shores of Spain." His confidence had, he explained, been founded on his knowledge of the agreement between the predictions made by Le Verrier and Adams. "The remarkable calculations of M. Le Verrier, if uncorroborated by repetition of the numerical calculations by another hand, or by independent investigation from another quarter, would hardly justify so strong an assurance as that conveyed by my expressions above alluded to," he declared. "But it was known to me, at that time (I will here take the liberty to cite the Astronomer-Royal as my authority), that a similar investigation had been independently entered into, and a conclusion coincident with M. Le Verrier's arrived at (in entire ignorance of his conclusions) by a young Cambridge mathematician, Mr Adams; who will, I hope, pardon this mention of his name (the matter being one of great historical moment) and who will, doubtless, in his own good time and manners, place his calculations before the public."

Challis decided to make details of his search public at the same time. Le Verrier's third paper had reached him at the end of September, and, unaware that the planet had already been discovered in Berlin, Challis decided to suspend his star mapping and to search for the planet by looking for its disk, just as Le Verrier suggested. On September 29, after examining 300 stars, he came upon one that seemed to have a disk, and made a note of its position. The following day, however, he learned that the planet had already been found, and that he had missed his chance of discovering it. He wrote immediately to the *Cambridge Chronicle*, and his letter was published on October 3. Challis explained about the agreement between the predictions made by Adams and Le Verrier, and then turned to his own attempts to find the planet. "Having been anticipated in the discovery of this planet, I need not detail the efforts I made to find it. I may, however, be permitted to state that for the last two months I have been engaged in mapping the stars in the neighbourhood of the probable place, which, though slow, must eventually have been successful."

Adams, in contrast, made no public statement about the discovery. He was bitterly disappointed that Galle had found the planet before Challis, but it was now too late to do anything about it. Instead, Adams saw the discovery as an opportunity for him to start a new set of calculations, and he immediately began calculating a more accurate orbit for the new planet, now that its exact position was known.

Airy arrived back in England on October 11. With Europe abuzz at the discovery, he had decided before his return that it would be a good idea to draw up a full account of the English side of the story. Herschel's letter to the *Athenaeum* had, after all, indicated that Airy had been fully aware of what was going on in Cambridge. With the existence of Adams's work and the failed Cambridge search now public knowledge, he was sure to be called upon to explain to the nation why, despite the head start provided by Adams, the planet had not been discovered in England.

The real reason, of course, was that although Challis had recorded the planet on two occasions, he had not analyzed his observations and had failed to realize what it was. Now that the planet had been found, Challis went back over his notebooks and found, to his horror, that the planet had been right under his nose all along. "I have been greatly mortified to find that my observations would have shewn me the planet in the early part of August, if I had only [examined] them," he wrote to Airy on October 12. "After four days of observing, the planet was in my grasp if only I had examined or mapped the observations." Challis explained that he had been overburdened with cometary calculations and had not had time to analyze his observations as he went along. "What is most provoking, I actually compared to a certain extent the observations of July 30 and August 12 soon after taking them . . . and for some unaccountable reason I stopped short within a very few stars of the Planet. . . . It is useless now to regret my having missed the planet when it was so possible to detect it. All that remains to do is to make the best of the observations that I have succeeded in getting." Challis added a final postscript: "We think at Cambridge that Oceanus would be a good name for the planet."

Clinging to the rather feeble triumph of having been the first to observe the planet, even though he had not realized it at the time, Challis wrote to the *Athenaeum*, whose pages were becoming the focus of the English discussion of the discovery. It must have been painful for Challis to make such a public acknowledgment of his failure to find the planet; but he put the best face on the matter that he could. "With respect to this remarkable discovery," he claimed, "English astronomers may lay claim to some merit." Challis explained that he had started looking for the planet in July, at the suggestion of the astronomer royal. And, he admitted, over the course of the first four nights of observing, he had recorded the position of the planet on two occasions: "Comparison of the observations of July 30 and August 12 would, according to the principle of research which I employed, have shown me the planet," he conceded. But, he added, he had not compared his observations until after the planet had been seen from Berlin. "The planet, however, was *secured*, and two positions of it recorded six weeks earlier here than in any other observatory, and in a systematic search expressly undertaken for that purpose." All of this, thought Challis, suggested that he and Adams had some claim to be able to name the new planet, and he advanced the name Oceanus "as one which may possibly receive the votes of astronomers."

By this time, the edition of the *Athenaeum* containing Sir John Herschel's letter had reached Paris, where it sparked an outcry. Who was this Adams, French astronomers asked, and why had he stayed silent until now? Le Verrier was furious. He took particular exception to Herschel's suggestion that a single calculation could not be trusted until it had been confirmed by a second, and began a heated correspondence with Airy that was to last several weeks. Le Verrier complained that Herschel was putting about the idea that his calculations alone were not worthy of confidence. "Oh! To merit the confidence of M. Herschel would undoubtedly be a great honour," he snarled. Since Airy had previously expressed such admiration for his work, Le Verrier hoped that Airy would leap to his defense in England.

So he was astonished to receive a letter from Airy in which his duplicity in failing to mention the work of Adams became apparent for the first time. Airy formally congratulated Le Verrier on the discovery and then continued, rather sheepishly: "I do not know whether you are aware that collateral researches had been going on in England, and that they had led to precisely the same result as yours. I think it probable that I shall be called on to give an account of these. If in this I shall give praise to others, I beg that you will not consider it as at all interfering with my acknowledgment of your claims. You are to be recognised, beyond doubt, as the real predictor of the planet's place. I may add that the English investigations, as I believe, were not quite so extensive as yours. They were known to me earlier than yours."

Le Verrier had been unaware of the existence of Adams's work, since Airy had singularly failed to mention it in his letter querying the radius vector. Airy caused further offense by suggesting in his letter that Oceanus might be a better name than Neptune, since it was a name "more similar in its character to that of its predecessor Uranus."

To French astronomers, it was clear what was going on: The English were trying to hijack the discovery. Challis, it turned out, had written to Arago and several other astronomers on October 5 detailing his sighting of the new planet on September 29, having looked for a disk, just as Le Verrier suggested; his letter had made no mention of Adams. Yet in England, he was now claiming to have seen the planet in August, using Adams's calculations. Le Verrier was baffled; Challis was, he complained in yet another angry letter to Airy, "saying 'white' in France, and 'black' in England."

The controversy surrounding the new planet dominated the October 19 meeting of the Academy of Sciences in Paris. In contrast to the triumphant tone of the October 5 meeting, the mood was one of outright fury. Arago gave a long and impassioned speech, defending Le Verrier, and attacking the "clandestine" work of Adams. "Le Verrier is called upon today to share the glory, so loyally, so rightly earned, with a young man who has communicated nothing to the public and whose calculations, more or less incomplete, are totally unknown in the observatories of Europe! No! No! The friends of science will not allow such a crying injustice to be perpetrated."

Arago quoted extensively from the letters of Airy, Herschel, and Challis, showing how inconsistent they were. Why did Airy have to ask Le Verrier about the radius vector if he already had Adams's predictions—why could he not have asked Adams? Why did Challis claim to be following the advice of Le Verrier one minute, and of Adams the next? "Mr Adams," Arago concluded, "has no right to figure in the history of the discovery of the planet Le Verrier, neither by a detailed citation, nor by the slightest allusion. In the eyes of every impartial man, this discovery will remain one of the most magnificent triumphs of astronomical theory, one of the glories of the Academy, and one of our country's noblest titles to the gratitude and admiration of posterity."

Le Verrier's cause was championed by *Le National* newspaper, in an article published on October 21 titled "A Planetary Theft," which accused Airy, Herschel, and Challis of organizing a "miserable plot" to rob Le Verrier of the honor of his discovery. The three men were, the newspaper noted by drawing on Arago's speech, incompetent thieves, since they had been tripped up by their own contradictory testimony. How convenient it was, the article went on, that Adams had supposedly solved the problem in October 1845, before Le Verrier had published anything on the subject—yet despite this, it had taken almost a year for the English to start looking for the planet. "There is no longer the slightest doubt," it continued sarcastically "that this is the young man who discovered the new planet, and, as he is too modest to name it Adamus, he will name it Oceanus; such is his right. . . . Now, how can Messrs Herschel, Airy and Challis possibly claim, in front of all of Europe, something so materially false as this? We cannot explain it." The authors of the article, who identified themselves only as "D and T," added that "this is certainly not a question of nationalism; science recognises no borders, any more than honour, or truth."

But there was, in reality, a great deal of nationalistic feeling behind the continued attacks by French newspapers on the English astronomers. *L'Univers* accused them of "an odious national jealousy," and *L'Illustration* published a series of cartoons of Adams. One of them showed him wearing a dunce's cap while peering through a telescope across the English Channel at Le Verrier's notebooks; another showed him looking through a telescope in the wrong direction altogether.

The French attacks were met with condescending ripostes in the English press. "We wish that the complete honour of this great fact had fallen to the English philosopher, but far beyond

French cartoons satirizing Adams. (Jean Loup Charmet, Paris)

any such national feeling is our desire that philosophers should recognise no such distinctions between themselves," declared the *Athenaeum.* "The petty jealousies of Earth are things too poor and mean to carry up among the stars. It is not essential to the recognition of Mr Le Verrier's merits that they should suffer Mr Adams' to be overlooked. The more valuable Mr Le Verrier's discovery, the more important it was that our English philosophers should show that they were on its track."

Adams himself, however, paid no attention to all of this international quarreling. His attitude was best summarized by a memorandum he had written to himself as a student: "Look without envy on the success of others, and without pride on my own."

It was not only in France that Challis and Herschel were criticized; the publication of their letters caused consternation in England as well, among astronomers who felt that they should have
been told of Adams's work. Most justified in this belief was John
Hind, who had been corresponding with Challis for several
months about the search for the new planet, only to discover in
October that Challis had been keeping quiet about Adams's results all along.

When Hind realized what had happened, he complained to
Richard Sheepshanks, a senior member of the Royal Astronomical Society, that he suspected the Cambridge astronomers of having kept Adams's work to themselves so that they could find the
planet first. "The Cambridge people do the best for their own,"
he wrote to Sheepshanks, lamenting that "I am sure you must

have noticed the inexcusable secrecy observed by all those acquainted with Mr Adams' results." Sheepshanks was only too
familiar with this secrecy because, as a Cambridge man himself,
he was one of the select few who knew of Adams's work.

The matter received a full public airing on November 13, at a
unusually crowded meeting of the Royal Astronomical Society in
London. Airy, Challis, and Adams presented papers detailing
their roles in the affair to a rapt audience that included many of
the country's leading scientific figures, as well as Adams's brothers George and Thomas. Adams's family and friends felt that he
had been let down by Airy and Challis, and George and Thomas
had come to see them being called to account for their failure.

Airy spoke first, to deliver a paper solemnly titled "Account
of Some Circumstances Historically Connected with the Discovery of the Planet Exterior to Uranus." The astronomer royal
started off by emphasizing the significance of the discovery. "In
the whole history of astronomy, I had almost said in the whole
history of science, there is nothing comparable to this," he declared. Although Le Verrier and Galle had been responsible for
the discovery, Airy went on, "we should do wrong if we considered that these two persons alone are to be regarded as the authors of the discovery of this planet. I am confident that it will
be found that the discovery is a consequence of what may prop-
erly be called a movement of the age; that it has been urged by
the feeling of the scientific world in general."

To emphasize this idea that the discovery was bound to have
happened sooner or later, and that concentrating too much on
the contribution of any one individual would be a mistake, Airy's
report was comprehensive, going all the way back to the speculation in the 1830s about the possibility of a planet beyond Uranus.

Airy recounted the whole story, with selected excerpts from the letters he had exchanged with Adams, Challis, and Le Verrier. He presented himself as an unbiased, impartial authority on the history of the events leading up to the discovery, since he had "not directly contributed either to the theoretical or to the observing parts of the discovery." The astronomer royal was, in other words, doing his best to distance himself from the controversy, implying that he had no direct hand in any of it and was therefore beyond blame or reproach.

Of course, Airy's account was not entirely candid. His reason, for example, for his failure to have replied to Le Verrier's letter explaining the radius vector query and offering a more accurate position for the planet was that "my approaching departure for the Continent made it useless for me to trouble M. Le Verrier with a request for the more accurate numbers to which he alludes." In fact, Airy had not set off for another six weeks; there had been plenty of time to write back to Le Verrier.

At the end of his account Airy drew several conclusions, all of which seemed primarily intended to defend himself. First, he made it clear that there was nothing unusual about paying less attention to a handwritten note than a memoir printed in an astronomical journal. He restated his conviction that the discovery of the planet was "a movement of the age." Finally, he concluded that in some cases, "the publication of theories, when so far matured as to leave no doubt of their general accuracy, should not be delayed till they are worked to the highest perfection" and suggested that had Adams published his results in October 1845, the planet might have been found soon afterward.

Challis was the next to speak, and his account made him look like an utter fool. He had been given Adams's prediction of the position of the planet even earlier than Airy but had failed to act upon it; when, almost a year later, he had finally mounted a proper search, he had seen the planet twice without realizing it and had then been beaten to the discovery by Galle. Even the Royal Astronomical Society's own historian subsequently described Challis's story as "pitiful . . . surely no feebler one was ever told. To do it justice it is candid. No one would dream of doubting its veracity, for what could induce any man to produce a tale of that complexion?"

Finally, Adams presented a summary of his method and calculations, concentrating on mathematical details rather than questions of priority or blame. He pointed out that his work had preceded that of Le Verrier, though he happily acknowledged Le Verrier's "just claims to the honours of the discovery, for there is no doubt that his researches were first published to the world, and led to the actual discovery of the planet by Dr Galle."

Adams was the only one of the three speakers to emerge from the Royal Astronomical Society meeting without looking foolish. In the weeks after the discovery, most English astronomers had regarded Le Verrier as the sole discoverer, since Adams was as unknown to them as he was to the French. But after hearing Adams explain his work in person, those present at the meeting could see that his calculations were not the vague scribblings that Airy and Arago had suggested but a detailed analysis of the problem that would have resulted in the discovery of the planet, had Challis not botched the search.

In the light of Adams's impressive account of his calculations, the extent to which Airy and Challis had failed to capitalize on his work became apparent. The two men were also accused of colluding to keep Adams's results secret. There was clearly some justification in this claim. Walter White, a member of the Royal Society who attended the meeting, wrote in his journal that it "appears from the delivery of Mr Airy's address that he is somewhat to blame for the loss of the complete realisation of the discovery of the new planet by Adams of Cambridge, and there appears to have been an attempt to make a Cambridge snuggery affair of it, for Challis and the Northumberland equatorial." Sir David Brewster, a Scottish scientist who was a fierce critic of Airy, made a similar suggestion, writing in the *North British Review*. He suggested that Challis had "striven to secure for Mr Adams the credit of his discovery, and to Cambridge the honour of having first detected the planet."

As well as being criticized by members of the Royal Astronomical Society, Airy faced sharp criticism from within Cambridge for his failure to act on Adams's prediction more quickly. Professor Adam Sedgwick of Trinity College, a geologist who was a friend of both Airy and Adams, was said to have cried out "Oh! Curse their narcotic souls" when he was informed over tea in the common room that Challis and Airy had allowed Galle to beat them to the discovery. Sedgwick summarized the feelings of many in a furious, almost illegible letter to Airy in which he accused the astronomer royal of "apathy" toward Adams. "Had the results communicated to you and Challis been sent to Berlin, I am told, they came so near the mark that to a certainty the new planet would have been made out . . . and the whole business settled in 1845—Adams the sole, unadvised, unassisted discoverer. Is this true? If so, I must chime in with the pack of grumblers. To say the least of it, a grand occasion has been thrown away," fumed Sedgwick, who, judging by his handwriting, was incoherent with rage. And when Le Verrier's second paper giving an estimate of the planet's longitude had appeared that summer, Sedgwick inquired, "Why in the name of wonder was not all Europe made to ring with the fact that a B.A. of Cambridge had done this ten months previously?"

In his reply, Airy tried once again to distance himself from the whole matter. "Those who in quarrels interpose, must often wipe a bloody nose," he remarked. He had, he explained, written to Adams to ask about the radius vector but had received no reply, which "entirely stopped me from writing again" and meant that, until Le Verrier's paper appeared, Airy lacked "sufficient grounds for trusting the results." As for publishing Adams's results, said Airy haughtily, that was hardly his concern; it should have been done by Adams himself or Challis.

In one of his many letters to Le Verrier, however, Airy conceded that Adams did have some right to feel let down by the lack of interest shown in his prediction. "You express yourself surprised that anyone should suppose that the results of mathematical investigations required confirmation," he wrote. "If any person had reason to complain of this it would be Mr Adams; for we waited until Mr Adams' results were confirmed by yours, and not until yours were confirmed by Mr Adams'. A sign may be taken wrongly, a decimal point may be placed in the wrong position, a 3 may be mistaken for an 8—there is no security whatever against such errors except by independent repetitions of the investigations."

Airy's continued efforts to present himself as nothing more than a minor player in the affair were further undermined by a conspiracy theory that started to circulate in December 1846.

Was it not possible, it was suggested, that Airy had been in league with Le Verrier? This implausible theory was enthusiastically propounded by an article in *Mechanics' Magazine* titled "Adams, the Discoverer of the New Planet." The author, who signed himself Exoniensis, declared that Adams's priority in having predicted the existence and position of the planet was beyond doubt, since he had delivered his results to Airy in October 1845. "It is idle and ridiculous then to raise any question respecting the priority of the discovery. At this point Mr Adams had no competitor; and had his astonishing achievements been duly followed up by the English astronomers to whom he made them known, there can be no doubt that the exterior planet would have been seen."

Why had Airy failed to act? The answer, the article suggested, was that he might have conspired with Le Verrier, passing him Adams's results and thus setting him on the trail of the planet. This, the author suggested, would explain why the two positions predicted for the planet were so similar; if Airy was conspiring with Le Verrier, it would also explain why he decided to trust Le Verrier's results but not those of Adams. "What was doubtful in English was accurate in French. . . . It is exceedingly painful and annoying to see Mr Adams, the undoubted pioneer and discoverer, slurred over, and his claims almost indistinctly alluded to, while Mr Le Verrier, who came into the field months after it was fully occupied by Mr Adams, gets the warmest laudation. Will the British public then quietly suffer one of its most gifted citizens to be deprived of his rights, and cajoled out of the results of his surprising labours? I trust not."

By the end of 1846, the controversy surrounding the discovery of the planet was still raging. Angry letters were flying between astronomers and illuminating the pages of newspapers and astro-

nomical journals. Scientists throughout Europe were at logger-
heads over the merit of Adams's claims, over who was to blame
for the English failure to find the planet first, and even over what
the planet should be called.

The international furor surrounding the new planet was in
marked contrast to the relatively straightforward nature of William Herschel's discovery of Uranus. Having died in 1822, William Herschel did not live to hear of the new planet that had
been found as the indirect result of his labors; but his sister Caroline did. At the age of ninety-six, she received the news of the
discovery in a letter from John Herschel's wife, Margaret.

Appropriately enough, given his father's indirect role in pre-
cipitating the discovery, John Herschel would prove instrumental
in resolving the dispute.

9

An Elegant Resolution

When Airy was told, he wouldn't believe it;
When Challis saw, he couldn't perceive it.

—CAMBRIDGE SLOGAN, 1846

Throughout the final months of 1846, there seemed little hope of brokering a peace. Sir John Herschel was appalled by the ferocity of the attacks being made on his name in France. "In bed half day after a sleepless night," he wrote in his diary on October 25, after being fiercely criticized by Le Verrier in a "savage" letter to the *Guardian*. As he considered his reply, however, Herschel was careful to strike as conciliatory a tone as possible.

In making John Couch Adams's work public, Herschel explained in his response, he had not meant to diminish Le Verrier's achievement. "I deeply regret that Mr Le Verrier should have found cause for complaint or offense in my communication to the Athenaeum," he wrote to the *Guardian*. "Nothing was ever further from my intention than to detract from the glory of his noble discovery, or tear one leaf from the wreath which he has so honourably won. The prize is his by all laws of fair adjudication, and there is not a man in England who will grudge him its possession." Indeed, Herschel noted, Adams himself "recognises M. Le Verrier's property in the discovery."

Ironically, by this time Herschel had realized that he had just missed discovering the planet himself a few years earlier. On July 14, 1830, he had been scrutinizing the part of the sky just half a degree from the planet's position at the time, using a powerful telescope that would have clearly shown its disk. But had he discovered the planet in this way, he decided, he would have robbed Adams and Le Verrier of the opportunity to distinguish themselves by deducing its existence mathematically. "It is better as it is," he wrote to a friend. "I should be sorry it had been detected by any accident or merely by its aspect. As it is, it a noble triumph for science." This highlights another reason why Herschel was so eager to end the dispute over the new planet: He believed that public bickering reflected badly on scientists, and that amid the name-calling and mudslinging, the significance of the discovery might be overlooked. Like many other Victorian scientists, Herschel saw the discovery as an opportunity to inspire the public with enthusiasm for science in general.

Herschel also extended an olive branch with respect to the planet's name, which was still the subject of much debate. While many astronomers on the continent had adopted the name Neptune, the French were insisting that the planet be called Le Ver-

John Herschel (*The Great Astronomers* by Robert S. Ball, London: Sir Isaac Pitman and Sons, 1895)

rier, and astronomers in England were divided. Aside from the fact that it would be a snub to Adams, calling the planet Le Verrier would also set a dangerous precedent by breaking with the mythological tradition. As W. H. Smyth, the president of the Royal Astronomical Society, remarked to George Airy, "Mythology is neutral ground. . . . Just think how awkward it would be if the next planet should be discovered by a German, by a Bugge, a Funk, or your hirsute friend Boguslawski!"

The name Oceanus, proposed by Challis with the backing of Adams, was only taken seriously within Cambridge. The names Chronos, Hyperion, Atlas, Atreus, and even Gravia (in homage to the theory of gravitation) had also been suggested by various astronomers but failed to win significant support. Since Le Verrier was now so opposed to the name Neptune (which he claimed to find "detestable") and would certainly not agree to Oceanus, Herschel suggested Minerva as a compromise. (Back in 1782, Minerva, along with Oceanus and Neptune, had been suggested as possible names for the planet Uranus.)

After the meeting of the Royal Astronomical Society on November 13, which further inflamed the situation, Herschel took additional steps to try to calm the storm. Richard Sheepshanks, a vocal supporter of Adams and Challis, had gone so far as to suggest that the latter ought to be recognized as the discoverer of the planet rather than Galle, since he had seen it first—even though he had not realized it at the time. Herschel argued that this was a ridiculous notion that would serve only to antagonize the French further. "Though Neptune ought to have been born an Englishman and a Cambridge man every inch of him," he wrote to Sheepshanks, it would be impossible to make "an English discovery of it, do what you will." He also suggested to Sheepshanks that the Royal Astronomical Society should try to avoid siding with either Adams or Le Verrier.

The Royal Society, which had awarded its annual Copley Medal to William Herschel in 1781 for his discovery of Uranus, wanted to honor Le Verrier in the same way, and it duly named Le Verrier as the recipient of the Copley Medal on November 30, 1846. By this time Herschel had made peace with Le Verrier, to the extent that Le Verrier even asked him to collect the Copley Medal on his behalf (a request that Herschel granted). Even so, Herschel believed that both Adams and Le Verrier deserved recognition, and suggested that the Royal Astronomical Society, which was considering the award of its annual Gold Medal, should give out two medals. The council of the Royal Astronomical Society was, however, deeply divided. Airy had nominated Adams, Challis, and Le Verrier for the medal to ensure that he appeared impartial, and several other astronomers had been nominated too. As a result, no candidate won a large enough share of votes to qualify for the medal. Throughout December, Herschel continued to press for the adoption of his compromise solution, in a steady stream of letters to members of the council. He also suggested that the wording of the commendations should avoid "the least allusion to that ugly word 'priority.'" Debate of the matter became so heated that at the end of one letter, to Sheepshanks, Herschel wrote "Burn this."

Airy was also doing his best to mend fences with his critics, having been "abused most savagely by both English and French," as he later complained in his autobiography. After a furious exchange of letters with Le Verrier, in which Airy complained that his private correspondence with Le Verrier had appeared in the French newspapers, the two men resolved their differences and resumed a cordial relationship. Indeed, the attacks in the French press had become so vociferous that Le Verrier and Arago both wrote to Airy to distance themselves formally from the remarks of their countrymen. The astronomer royal also made his peace with Challis, after a slightly strained correspondence in which each accused the other of trying to pass the buck in their official accounts.

As for Adams, who had taken no part in the public argument over priority, he wrote to Airy to apologize for his failure to respond to the radius vector query. He had not realized, he explained, "the importance which you attached to my answer on this point," though he admitted that he had been "much pained at not having been able to see you when I called at the Royal Observatory the second time, as I felt that the whole matter might be better explained by half an hour's conversation than by several letters." He too was soon corresponding happily with Airy and was evidently far more interested in discussing the orbital characteristics of the new planet than the circumstances of its discovery.

With his passion for order, Airy bound all of these letters, along with Adams's original note predicting the new planet's position, a sheaf of newspaper clippings, and all other documents relating to the matter, into a scrapbook. He labeled it "Papers relating to the Discovery, Observations and Elements of Neptune" and filed it away in the archives of the Royal Observatory.

By the beginning of 1847, the astronomers of Europe had all the facts at their disposal. Le Verrier's papers had been printed the previous year, and Adams's paper detailing his calculations had appeared as a special supplement to the *British Nautical Almanac.* The accounts that Airy, Challis, and Adams had read to the Royal Astronomical Society had also been widely distributed. As a result, a consensus started to form over the still-unresolved questions of naming and priority. In particular, the elegance of Adams's paper convinced many astronomers that he was indeed entitled to some claim in the discovery.

The Danish astronomer Peter Hansen, who was generally regarded as the world's leading expert on gravitational perturbations, declared that he regarded Adams's paper as more mathematically beautiful than Le Verrier's. He was not alone.

Writing to the French astronomer Jean Baptiste Biot, Airy admitted that "on the whole, I think his mathematical investigation superior to M. Le Verrier's. However, both are so admirable that it is difficult to say." In Paris, Biot had written a conciliatory article in which he conceded that Adams had deduced the existence of the new planet before Le Verrier, even though he had not published his results. "I only see a talented young man who has been served badly by circumstances this one time, and whom one must applaud in spite of it. I say to him: the laurel which you merited first, another merited also, and has carried it away before you had the boldness to seize it," he declared.

Herschel was still trying to persuade the Royal Astronomical Society to change its rules and award two medals, since this would have a "healing and friendly effect." At a second meeting of the council to decide who should be awarded the Gold Medal, the field was narrowed to two candidates, Adams and Le Verrier. But neither received the necessary twelve votes, because five of the council's fifteen members now backed Adams.

Despite the deadlock at the Royal Astronomical Society, Herschel's suggestion that both men were entitled to equal credit was rapidly gaining ground elsewhere. The validity of Adams's claim had been recognized by Russian astronomers at a meeting of the Imperial Academy of Sciences in St. Petersburg. As a result, the Russians decided, Neptune was the most appropriate name for the new planet, since "the name of Le Verrier would be against the accepted analogy, and against historical truth, as it cannot be denied that Mr Adams has been the first theoretical discoverer of that body, though not so happy as to effect a direct result of his indications."

Elsewhere in Europe, the name Neptune had become the

clear favorite. Heinrich Schumacher, whose opinion as the editor of the *Astronomische Nachrichten* carried a lot of weight, wrote to Airy to complain of Arago's arrogance in trying to foist the name Le Verrier on the astronomical community. By and large, he explained, German astronomers (including Gauss and Encke) had already adopted Neptune. Airy made one final attempt to encourage Le Verrier to suggest an alternative mythological name, but Le Verrier refused to do so, so on February 28, 1846, Airy reluctantly announced that he had settled on Neptune as well; Challis and Adams swiftly fell in line with his decision.

With the name of the planet resolved (outside France, at any rate), John Herschel continued to do his best to win some form of official recognition for Adams. In addition to the Copley Medal, Le Verrier had received several other honors following his discovery. He had been made an officer in the Legion of Honour and preceptor to the count of Paris; a special chair in astronomy had been created for him at the Faculty of Sciences in Paris; the grand duke of Tuscany honored him with a new edition of the works of Galileo; the king of Denmark awarded him the order of the Danneborg; and he was granted honorary membership of the Imperial Academy of Sciences in St. Petersburg and the Royal Society of Göttingen. Adams, on the other hand, was still merely an unrecognized graduate student. But in June 1847, thanks to the combined influence of Herschel and Sedgwick, the twenty-eight-year-old Adams was offered a form of recognition that his supporters felt reflected the magnitude of his achievements: a knighthood from Queen Victoria.

Adams was honored to have been offered the same title that had previously been bestowed upon his hero, Isaac Newton. Yet at the same time, as he explained in a letter to Sedgwick (through whom the offer had been made), "it is doubtful whether I could afford to accept such an honour." Since he had no private fortune, Adams explained, he would still be dependent on private teaching to supplement his income, which would seem out of keeping with his title; it might also present difficulties if he wanted to get married, since "my choice would be seriously restricted by the necessity there would be for keeping up appearances." Finally, Adams noted, "my father is simply a farmer, and it might appear rather incongruous that his son should be Sir John." With characteristic modesty, Adams turned down the knighthood.

By the summer of 1847 the orbit of Neptune had been determined, and the nature of the planet's influence on Uranus could at last be fully fathomed. Neptune's orbit was calculated, like that of Uranus before it, with the help of historical data. Just as they had searched for prediscovery observations of Uranus in order to make calculation of its orbit easier, astronomers looked for occasions on which Neptune had been observed but mistaken for a star. In February 1847 just such an observation was found by the American astronomer Sears Cook Walker at the National Observatory in Washington, D.C. (The discovery of Neptune had been greeted with great enthusiasm in the United States; *Scientific American* called it "perhaps the greatest triumph of science upon record." The ensuing controversy was also keenly followed in the press.)

Walker found that half a century earlier, on May 10, 1795, the French astronomer Michel Lalande had observed Neptune and taken it for a star. Subsequent examination of Lalande's original observing notebooks revealed that he had also seen the planet on May 8 and noticed that it had moved two days later. But rather than concluding that he had seen an undiscovered planet, Lalande simply assumed that his first observation had been inaccurate and thus missed the opportunity of making a great discovery. Walker remarked that "one more wonder was added to the list of strange events in the history of Neptune." When news of Lalande's prediscovery observation reached Airy, he remarked, "Let no one after this blame Challis."

In Europe, Lalande's observation was independently discovered by Adolf Cornelius Petersen at the Altona Observatory in Germany. Before long, several astronomers, including Adams and Le Verrier themselves, had used the 1795 sighting to calculate reasonably accurate orbits for Neptune. And by comparing the relative positions of Uranus and Neptune over the years, they could see why Uranus had behaved in the way it did, and why it had baffled two generations of astronomers.

Throughout most of the eighteenth century, it turned out, Neptune had been far away from Uranus, and the gravitational forces between the two planets had been all but negligible. But by the beginning of the nineteenth century, they were moving closer together; they were closest in 1822, when Uranus overtook the slower-moving Neptune on its way around the Sun. The effect of Neptune's gravitational influence was to cause Uranus to move more quickly than it otherwise would have between 1800 and 1822, and more slowly between 1822 and 1840.

This explained why Alexis Bouvard had been unable to reconcile the ancient (pre-1781) observations with the modern (1781–1820) ones in his 1821 tables. During the period covered by the ancient observations, the motion of Uranus was hardly affected

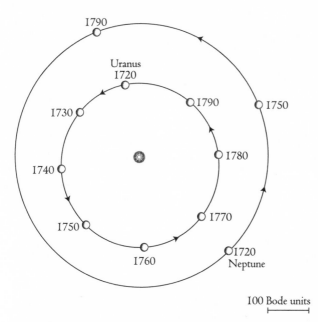

The positions of Uranus and Neptune between 1720 and 1790. The planets were far apart during this period, so the effect of gravitational perturbations was negligible.

at all by the influence of Neptune; but shortly after its discovery, Uranus started to succumb to Neptune's gentle tugging. Bouvard chose to ignore the ancient observations and base his tables on modern observations, which covered a period in which Uranus was, in fact, speeding up. But just after Bouvard's tables appeared in print, Uranus overtook Neptune and began to slow down. So it is hardly surprising that the tables were so flawed.

Although the mystery of Uranus's motion finally appeared to have been solved, the determination of Neptune's orbit raised another question. It was immediately apparent that the orbit of

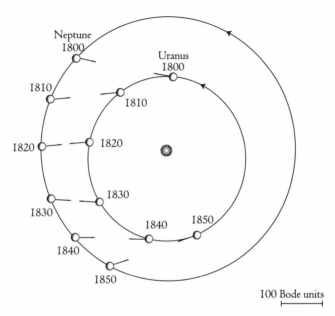

The positions of Uranus and Neptune between 1800 and 1850. The effect of Neptune's gravitational influence was to cause Uranus to speed up between 1800 and 1822, and slow down between 1822 and 1840.

Neptune was very different from the orbits calculated by Adams and Le Verrier. In particular, the planet's average distance from the sun was found to be around 303 units on the Bode scale, which was much less than the 388 units predicted by Bode's law, which both Adams and Le Verrier had used as the basis of their initial calculations. Admittedly, both of them had subsequently concluded that a smaller value was likely; Le Verrier's third paper calculated the average distance as 362 units, and Adams's second hypothesis put it at 373 units. So while the discovery of Neptune had triumphantly vindicated Newton's law of gravitation, it also dealt a mortal blow to Bode's law.

Planet	Radius of Orbit	Predicted Radius
Mercury	4	4 + 0 = 4
Venus	7	4 + (1 × 3) = 7
Earth	10	4 + (2 × 3) = 10
Mars	15	4 + (4 × 3) = 16
asteroids Ceres, Pallas, etc.	28	4 + (8 × 3) = 28
Jupiter	52	4 + (16 × 3) = 52
Saturn	95	4 + (32 × 3) = 100
Uranus	192	4 + (64 × 3) = 196
Neptune	303	4 + (128 × 3) = 388

The differences between the orbits of the postulated planets and of Neptune itself led the American astronomer Benjamin Pierce to make a stunning claim. Pierce, the professor of astronomy and mathematics at Harvard, had a reputation for stirring up controversy. One contemporary described him as "hot-tempered and hasty," and on this occasion, at least, it seems that he got rather carried away. On March 16, 1847, he told a meeting of the American Academy of Arts and Sciences in Boston that "the planet Neptune is not the planet to which geometrical analysis had directed the telescope. . . . its discovery by Galle must be regarded as a happy accident."

Pierce insisted that he had nothing but the utmost admiration for Le Verrier's work. "I have studied his writings with infinite delight, and am ready to unite with the whole world in doing homage to him as the first geometer of the age, and as founder of a wholly new department of 'Invisible Astronomy,' " he said. But he was convinced that Neptune "is not the planet of Le Verrier's or Adams's theory."

The basis of Pierce's statement is clear when the actual orbit of Neptune and the orbits of the planets predicted by Le Verrier and Adams are plotted on a diagram. It is obvious that the actual and calculated orbits are very different. Neptune's orbit is almost circular, whereas the orbits calculated by Le Verrier and Adams are noticeably elliptical.

How, then, had such apparently inaccurate orbits been able

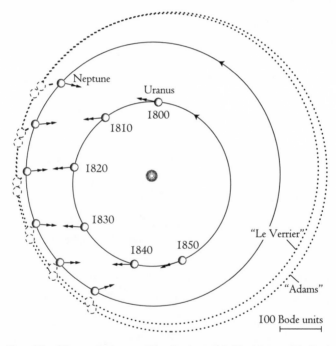

The orbits of Uranus, Neptune, and the planets predicted by Adams and Le Verrier. Although the predicted planets do not have the same orbital characteristics as Neptune, during the crucial period of 1800 to 1850 their gravitational influence on Uranus would have been very similar to that of Neptune.

to predict Neptune's position? After months of debate and analysis, mathematicians on both sides of the Atlantic determined the answer. They were helped by the discovery of a moon of Neptune, which was first seen by the English amateur astronomer William Lassell on October 10, 1846, but whose existence was not confirmed until the following July. From observations of this moon, astronomers were able to calculate Neptune's mass, which was found to be roughly equal to that of Uranus. Neptune was, in other words, about half as massive as Adams and Le Verrier had thought it was. But it was also closer to the Sun (and hence to Uranus) than expected. As a result, during the crucial period of 1800 to 1850, when Uranus was being influenced by the new planet, the gravitational effect of each of the predicted planets turned out to be almost exactly the same as the actual effect of Neptune. The predicted planets' greater mass had compensated for their greater distance.

Although some American astronomers continued to insist that the prediction of Neptune's position had been nothing more than a lucky fluke, most astronomers were happy with this explanation. After all, as John Herschel later noted in his book *Outlines of Astronomy,* published in 1850, the task Adams and Le Verrier had embarked upon was "not to determine as astronomical quantities the axis, eccentricity and mass of the disturbing planet, but practically to discover where to look for it." And in that, they had succeeded.

After two years of argument, debate, and mudslinging, Adams and Le Verrier met for the first time in June 1847 at a meeting of the British Association in Oxford. Neither of the two men had ever cast aspersions on the other during the various arguments over the new planet. Even so, the other astronomers present were unsure how they would react to one another. They need not have worried; Adams and Le Verrier cordially and sincerely shook hands, to the delight of everyone present, and before long they were chatting away like old friends. A few days later John Herschel hosted a private party at his country house, to which both men were invited. Astonishingly, given their different temperaments, but perhaps understandably, given their mutual interests, they struck up a friendship that was to endure for the rest of their lives.

The following year, having been elected president of the Royal Astronomical Society, John Herschel was finally able to resolve the fiasco of the Gold Medal. Since the society's rules forbade the awarding of more than one medal, special "testimonials" were awarded to Adams and Le Verrier instead. Herschel made a florid speech in which he publicly spelled out his belief that the two men should be regarded as equals. The names Le Verrier and Adams were, he said, "names which, as Genius and Destiny have joined them, I shall by no means put asunder; nor will they ever be pronounced apart so long as language shall celebrate the triumphs of Science in her sublimest walks. . . . As they have met as brothers, and as such will, I trust, ever regard each other, we have made—we could make—no distinction between them on this occasion. May they both long adorn and augment our science, and add to their own fame, already so high and so pure, by fresh achievement!"

In Neptune's Sway

Dear old Georgium Sidus,
I shan't forget what pains he took to hide us.
His fidgetting, at last, awoke suspicion,
And pointed out exactly my position.

—J. R. PLANCHE
FROM *THE NEW PLANET . . . AN EXTRAVAGANZA IN
ONE ACT*

Not only did the planet Neptune affect the course of Uranus
through the heavens; it also determined the paths of the subsequent careers of those involved in its discovery. The chief bene-
ficiary was Urbain Jean-Joseph Le Verrier. Soon after the
discovery he was asked by the French government to draw up a

national plan for future astronomical research. Le Verrier conceived a grand scheme which would take many years of effort: to recalculate from scratch entirely new tables of the motions of the planets and embrace the entire planetary system in a single work. With characteristic self-confidence, Le Verrier based his plan, which he delivered in February 1847, on the assumption that he would soon be appointed in Arago's place as director of the Paris Observatory, the most senior post in French astronomy.

It was, however, several years before Le Verrier was able to put his plan into action. Only in 1854, following Arago's death, did Le Verrier succeed him as director—an appointment that was a foregone conclusion given Le Verrier's ability, fame, and growing political influence. He had been elected to France's Legislative Assembly in 1847, was made a senator in 1852, and was a member of the committee appointed to choose Arago's successor. Le Verrier was installed as director of the observatory on January 31, 1854.

Although Le Verrier was a gifted theorist with little experience in practical astronomy, he quickly acquainted himself with the subject and instituted a thorough reform of the observatory's instruments and procedures. But his authoritarian character and haughty, egotistical nature meant that he soon alienated himself from his staff. One of Le Verrier's new rules stipulated, for example, that there was no need to make public the names of assistant astronomers who made discoveries, "when all the merit is due exclusively to the director whose orders they were following . . . besides, these young astronomers receive a bonus and a medal for each discovery." The observatory's employees soon realized that the new director was not a man to be trifled with. One British writer who visited the Paris Observatory shortly after Le

Verrier's appointment described being "the unwilling witness of a painful scene, when an old assistant, who had been many years employed at the Observatory under Arago, was peremptorily discharged for a very trivial breach of duty."

In many respects, Le Verrier ran the Paris Observatory in a manner similar to the way Airy ran the Greenwich Observatory. Assistants were regarded as mere drudges; telescopes and quadrants were only means of procuring data for theoretical analysis. Like Airy, Le Verrier took little part in making observations himself. One of his contemporaries, Camille Flammarion, went so far as to suggest that Le Verrier was not even interested in observing the planet Neptune through a telescope. "I believe that he never saw it," Flammarion wrote in 1894. "For him, astronomy was entirely enclosed in formulae—I asked him if he thought that the other planets might be inhabited like ours, what might be the retinue of innumerable suns scattered in immensity, what astonishing coloured lights the double stars should shed on the unknown planets which gravitate in these distant systems. His replies always showed me that these questions had no interest for him, and that in his opinion the essential knowledge of the universe consisted in equations, formulae and logarithmic series."

Having established his authority at the Paris Observatory, Le Verrier began the execution of his scheme to draw up an entirely new set of planetary tables. He started with the inner planets: Mercury, Venus, Earth, and Mars, whose orbits he had considered back in the 1840s prior to his investigation of Uranus. Le Verrier was particularly interested in the motion of the planet

Mercury, whose orbit did not quite seem to accord with the predictions of gravitational theory.

Unlike the orbit of Uranus, whose motion had been entirely unpredictable until the discovery of Neptune, Mercury's was anomalous in a predictable way. Observations going back hundreds of years showed that the ellipse traced out by Mercury as it moved around the Sun was slowly rotating by 565 arcseconds per century, as a result of gravitational perturbation by other planets, primarily Venus. But when Le Verrier calculated the rate of rotation that would be expected from theory, he found that it was just 527 arcseconds per century. Only by including an arbitrary correction of 38 arcseconds per century could the planet's motion be predicted accurately.

Such an anomaly could not be swept under the carpet by blaming it on observational error, since the position of Mercury can be measured extremely accurately during a "transit," when

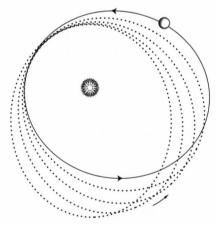

The slow pivoting of Mercury's orbit, grossly exaggerated.

the planet passes between the Earth and the Sun and appears as a black dot that moves across the Sun's disk. After carrying out a detailed analysis of Mercury's motion based on 411 observations, Le Verrier concluded that there was definitely something amiss. Perhaps unsurprisingly, given his successful prediction of the existence and position of Neptune, he started to consider the possibility of a new planet, orbiting between Mercury and the Sun, and responsible for perturbing Mercury in such a way as to compensate for the discrepancy between observation and theory.

This was not a new idea; a number of other astronomers had already proposed that such a planet might exist. But there was a major problem with the hypothesis. Although a planet orbiting so close to the Sun would normally be difficult to observe due to the Sun's glare, it ought to be strikingly visible during a total eclipse of the Sun; yet no such planet had ever been seen. So Le Verrier considered the possibility that there might be several smaller bodies, forming a second asteroid belt, within the orbit of Mercury. Each of these bodies would be small enough to have remained unnoticed, but collectively they would explain the slow swiveling of Mercury's orbit. Some of these bodies, he suggested in a paper on the subject, might be large enough to be seen as they passed in front of the Sun, when they would appear as small black spots crossing the Sun's disk. "This should make astronomers the more zealous to study the Sun's surface every day. It is important that every regular spot appearing on the Sun's disc, however tiny, be followed for a few months to determine its nature," he wrote.

Soon after Le Verrier published his analysis of the Mercury problem, he received a letter from a French amateur astronomer and rural doctor called Edmond Lescarbault, who claimed to have seen a black spot moving across the Sun on March 26, 1859. Exhilarated at the prospect of having discovered a new world, Lescarbault gave a detailed account of his observations and noted that "the planet appears as a black dot with a well-defined circular perimeter . . . this body is the planet, or one of the planets, whose existence you, Monsieur Directeur, recently revealed near the Sun through your marvellously persuasive calculations; these are the same which in 1846 enabled you to recognise Neptune's existence, to fix its position and to trace its course through the depths of space."

Unfortunately for Lescarbault and Le Verrier, however, this was not to be a rerun of the Neptune discovery. After rushing to Lescarbault's village to interrogate the doctor and make sure his observations could be trusted, Le Verrier worked out the orbit of the body, which was swiftly named Vulcan by another enterprising Frenchman, and arranged for Lescarbault to receive the award of the Legion of Honour. But despite Le Verrier's efforts to predict its position, the phantom planet was never seen again. Worse, Emmanuel Liais, astronomer to the emperor of Brazil, who had been observing the Sun at the same time as Lescarbault but with a more powerful telescope, said that he had not seen anything. Even so, Le Verrier believed in the existence of Vulcan for the rest of his life and made several further attempts to predict its position, none of which succeeded. (Mercury's anomalous behavior was not explained until 1915, when Albert Einstein published his theory of general relativity, which showed that Newton's law of gravitation requires a slight modification for planets as close to the Sun as Mercury.)

Meanwhile, work continued on Le Verrier's new set of planetary tables. In January 1870, however, the staff of the Paris Observatory decided that they had had enough of Le Verrier's despotic behavior and threatened a walkout. This was merely the most dramatic in a long series of disagreements between Le Verrier and his subordinates, and, following a government inquiry, Le Verrier was sacked. He was replaced by Charles Delaunay, who ran the observatory for two years but was drowned in a boating accident in 1872. Le Verrier was then reinstated under the supervision of a special council to ensure he behaved himself.

In 1876 Le Verrier received a medal from the Royal Astronomical Society in Britain in recognition of the value of his new tables, which were by this time nearing completion. John Couch Adams gave an eloquent speech praising Le Verrier's mathematical abilities and expressing "the interest with which we have followed his unwearied researches, and the admiration which we feel for the skill and perseverance by which he has succeeded in binding all the principal planets of our system, from Mercury to Neptune, in the chains of his analysis." Adams looked forward to the publication of the last few volumes and expressed the hope that Le Verrier would "thus be able to rest awhile and re-establish his health, shaken, we fear, by his too arduous labours—until he goes forth again, with fresh vigour, to win new triumphs in the fields of Physical Astronomy."

Because of poor health, Le Verrier was not present to hear these words and receive his medal. One of his last acts was to sign off on the final pages of his great work: the tables, appropriately enough, relating to the planet Neptune. With his life's work completed, Le Verrier died in Paris on September 23, 1877, exactly thirty-one years to the day after the discovery that had made his name. As a grand officer of the Legion of Honour, he was buried with full military honors, and his pallbearers included several leading astronomers. On June 27, 1889, a statue of Le Verrier was unveiled in front of the Paris Observatory. And when Alexandre-Gustave Eiffel completed his famous Parisian tower that same year, he included Le Verrier among seventy-two prominent French scientists whose names were inscribed on plaques around the first stage.

John Couch Adams also prospered as a result of the discovery of Neptune. In 1848 he was belatedly awarded the Royal Society's Copley Medal in recognition of his role in the affair and was thus formally recognized as the equal of Le Verrier, who had received the same award in 1846. But in contrast to Le Verrier, who was politically active and who spent the last years of his life subjugating the solar system through mathematical analysis, Adams chose the quiet life of an academic. In 1858 he became professor of astronomy and geometry at Cambridge, and in 1861 he succeeded James Challis as the director of the Cambridge Observatory.

As a senior Cambridge academic and a gifted theorist, Adams earned a reputation for setting examination questions of particular mathematical elegance. But he was also criticized for his inability to prevent his students from cheating. "Surely a man who discovers planets can find some practical way of conducting an examination fairly, and stop the low system of copying," complained one of his contemporaries. Adams's astronomical research, meanwhile, was chiefly concerned with the motion of the Moon, and the slow enlargement of its orbit around the Earth, which he placed on a firmer footing than anyone before him and thus earned a second Copley Medal from the Royal Society in 1866. He also did valuable research into the nature of the Earth's magnetic field and into the relationship between meteor showers and comets.

At around the same time that Le Verrier was searching for Vulcan, Adams discovered a new world of his own: He got married. In a letter to his friend, the geologist Adam Sedgwick (who had suggested to Adams that, at the age of forty, he ought to hurry up and settle down), he wrote: "I have followed your advice, have boldly put the important question, and have been rewarded with a favourable reply . . . already I feel as if in a new world, and look back with pity on my former state as on a glacial period, removed from the present by long geological ages."

Adams was twice elected president of the Royal Astronomical Society, and in that capacity he awarded prizes to both d'Arrest and Le Verrier for their contributions to the progress of astronomy. When Isaac Newton's papers were bequeathed to Cambridge University in 1872, Adams was asked to edit and catalog them. He was also the first president of the Association for Promoting the Higher Education of Women in Cambridge, was among the first professors to open his lectures to women, and was involved in the establishment of Newnham College for female students in 1880. His status as the leading English astronomer of his generation was confirmed in 1881, when George Airy retired as astronomer royal, and Adams was offered England's most senior astronomical post. He modestly declined the offer, on the grounds of his age (he was sixty-two).

In October 1884 Adams represented Britain at the international conference convened in Washington, D.C., to decide on a single global standard for the measurement of terrestrial longitude. At the time, maps and charts measured longitude from a variety of meridians: British charts used the Greenwich meridian as 0° longitude, French charts used the meridian in Paris, and other maps used meridians in Cadiz, Naples, Lisbon, Stockholm, and various other places around the world. Agreeing on a single meridian, to be internationally recognized as longitude zero, would make life easier for surveyors and navigators, who would no longer have to convert between coordinate systems. Eventually, despite the objections of the French delegate, the Greenwich meridian laid down by George Airy in 1851 (one of several meridians ruled by astronomers royal at Greenwich) was chosen on the practical grounds that it was the most widely used. Adams thus secured Airy's immortality as the ruler of the prime meridian.

Adams continued to live at Cambridge Observatory, the site of Challis's failed search for Neptune, until the end of his life. In addition to his astronomical work, he had a keen interest in botany, geology, and history. As a form of relaxation, he enjoyed calculating the values of mathematical constants to more than 200 decimal places. For Adams, the appeal of this apparently pointless exercise lay in the challenge of manipulating extremely long strings of figures without error—the mathematical equivalent of keeping several plates spinning in the air at once, or juggling with a large number of balls. Such esoteric pastimes aside, he also liked to play croquet, bowls, and whist. At the end of a long and distinguished academic career, Adams died on January 21, 1892, at the age of seventy-two.

George Airy spent the years after his retirement in 1881 working on his own calculations on the motion of the Moon, and some of his results were published in 1886. Comparison with observations made by Delaunay, however, revealed that Airy's calculations contained a hidden flaw. "I cannot but fear that the error which is the source of discordance must be on my part," Airy noted, and spent the rest of his life searching for the mistake. But he was unable able to pin down the error by the time of his death on January 2, 1892, at the age of ninety. A tragic note was later found among the papers of his flawed calculations: "When I discovered that I had committed a grievous error in the first stage of giving numerical value to the theory . . . my spirit in the work was broken, and I have never heartily proceeded with it since."

With Adams and Airy having died within a few days of each other, the British astronomical community immediately began to consider how best to honor two such eminent astronomers. In Cambridge, a meeting was held at St. John's College at which it was decided that a memorial to Adams should be placed in Westminster Abbey. And so in 1895, fifty years after Adams had delivered his prediction of Neptune's position to Airy, a plaque was ceremonially unveiled in the Abbey near the graves of William Herschel, Charles Darwin, and Isaac Newton, inscribed with the Latin words "JOHANNES COUCH ADAMS, Planetam Neptunum Calculo Monstravit MDCCCXLV" (John Couch Adams, pointed out the planet Neptune through calculation in 1845).

Those present, who included many of Britain's leading scientists, hailed Adams for his mathematical accomplishments, his modesty, and his tact in having abstained from the controversy surrounding the discovery of Neptune. Lord Kelvin (after whom the unit of temperature used by scientists was subsequently named) made a speech describing Adams as Newton's successor, and particularly commended Adams's work on lunar theory as even more of an accomplishment than his Neptune calculations. In addition to the plaque, a bust of Adams was commissioned, and today it is one of two statues adorning the main staircase at the headquarters of the Royal Astronomical Society in Piccadilly; the other is of Adams's hero, Newton himself.

There was to be no plaque in Westminster Abbey for George Airy, however. Even five decades after the Neptune affair, there was still much bad feeling toward the late astronomer royal. As one astronomer noted in a letter to Airy's successor, William Christie, the fact "that Cambridge should have been furious about the matter is easy to understand, all England was more or less, and the call for someone to hang was met by fixing on Airy and, strange to say, excluding Challis; but I should have thought that people now were calmer. Unfortunately, they are not. I do not suppose that Airy will have very many zealous supporters. It is the man's work and not himself that will attract them: I fancy that to the world at large he must always have seemed cold and repelling." The idea of honoring Airy with a plaque gained little support and was swiftly forgotten.

The Neptune controversy had, in fact, continued to rumble for many years. Shortly before his death in 1854, Arago published his account of the affair in a book called *Popular Astronomy*. He maintained his position that Adams had "no right to participate in the glory of the discovery of Neptune" since he had not published his calculations. When the book was translated into English, however, the translator added an angry footnote—"These remarks contain several mis-statements which it is desirable to correct"—and went on to defend Adams at some length.

In 1896, on the occasion of the fiftieth anniversary of the discovery, the English astronomer Sir Robert Ball wrote an article recounting the tale in *Strand* magazine. He criticized "certain French writers" for their failure to credit Adams, adding that "impartial judges generally refer to it as the joint result of the concurrent labours of the French and English astronomers." In 1905 William Ellis, a former assistant astronomer under Airy, wrote a letter to the journal *Observatory* defending the conduct of his former employer during the Neptune affair. "The feeling that was aroused at the time seemed unduly to overshadow him for the remainder of his life," he complained. The journal's editor, anxious to avoid reopening old wounds, published the letter but added a note declaring that "we do not wish any further correspondence on this subject."

Finally in 1911, after the death of Galle, his obituarist at the Royal Astronomical Society, Herbert Hall Turner, made a valiant attempt to draw a line under the affair. "With the death of Johann Gottfried Galle the curtain finally descends upon the great drama of the discovery of Neptune, the opening scenes of which thrilled the whole world two-thirds of a century ago," he wrote. "The end is calm after storm. The exit of the last actor stirs no memories of conflict save by association, for Galle took no part in the scenes where the battle raged." Within a few paragraphs, however, Turner had started speculating about what might have happened if Galle had been denied permission to search for Neptune, and Challis had examined his observations more closely.

The participants in the Neptune story left an enduring astronomical controversy behind them. More important, however, they also bequeathed to posterity a valuable new way to discover planets. No longer would astronomers search for planets by randomly peering through telescopes; instead, they would engage in complex calculations in order to detect the presence of worlds unseen. Who would be the first to emulate the success of Adams and Le Verrier by discovering a new planet? The race was on.

11

Shots in the Dark

*Forever invisible to the unaided eye of man, a sister-globe to our earth was shown
to circulate, in frozen exile, at 30 times its distance from the sun. Nay, the possibility
was made apparent that the limits of our system were not even thus reached, but that
yet profounder abysses of space might shelter obedient, though little favoured
members of the solar family, by future astronomers to be recognised through the
sympathetic thrillings of Neptune, even as Neptune himself was recognised through
the tell-tale deviations of Uranus.*

—Agnes Clerke
*A Popular History of Astronomy during the
Nineteenth Century,* 1886

"The discovery of a new planet in a new way, by first finding
where a planet ought to be, has given a fresh impulse to the

enthusiasm of astronomers," declared *Scientific American* in March 1847. "All are looking to see if the motion of heavenly bodies in some other direction does not indicate that there are more weights in the scale on that side than have yet been seen."

Urbain Jean-Joseph Le Verrier himself had suggested that the technique used to reveal Neptune could lead to further discoveries. "This success," he wrote shortly after the discovery, "must allow us to hope that after 30 or 40 years of observations of the new planet, it will be possible to use it, in turn, to discover the next planet in order of distance from the sun. And then the next; soon the planets will, unfortunately, become invisible, due to their immense distance from the sun, but their orbits will, in the following centuries, be worked out with great accuracy, through the use of theory."

By the time of Le Verrier's death in 1877, Neptune's mass and orbit had been established with reasonable accuracy, and a number of astronomers began to speculate that the new planet was unable to account for the anomalous motion of Uranus, after all. For when the gravitational effects of Neptune were taken into account, it seemed there was still a tiny discrepancy between the predicted and observed positions of Uranus. Was another unseen planet to blame?

In 1880 George Forbes, a Scottish astronomer, predicted the existence of a planet, larger than Jupiter, at a distance of 100 astronomical units. (An astronomical unit is the average distance from the Earth to the Sun, a convenient astronomical yardstick that is equal to 10 units on the Bode scale.) By this time a new technology had become available to planet-hunting astronomers: astro-photography. By taking a photograph through a telescope of a particular region of the sky, and then taking another picture of the same region a few days later and comparing the two, an astronomer could distinguish moving planets from stationary stars. In effect, photography automated the tedious process of star mapping that Challis had used to search for Neptune. A photographic search for the new planet predicted by Forbes, however, proved fruitless.

Forbes then returned to the problem with a new prediction, derived from observations of comets whose orbits he suspected had been altered by an unseen planet. This time he put its distance at 105 astronomical units (AU) and suggested that it moved in a highly tilted orbit that carried it far outside the zodiac—which would explain why nobody had noticed it so far, since most planet hunters restricted their searches to the region defined by the constellations of the zodiac.

Similarly, the existence of other planets was predicted by the American astronomer David Peck Todd, who mounted several futile searches. Thomas Jefferson See, an eccentric American astronomer, claimed on the basis of a dubious theory of his own devising that there were three planets beyond Neptune, orbiting the Sun at distances of 42, 56, and 72 AU. American astronomer William Pickering proposed a planet with twice the mass of Earth, orbiting at a distance of 52 AU. Among the dozens of predictions, perhaps the grandest claim of all was the assertion made by a Russian amateur astronomer, General Alexander Garnowsky. In 1902, in a letter to a French astronomical journal, he announced that he had calculated that there were no fewer than four new planets beyond Neptune. When asked for further details, he failed to reply. Neither his planets nor those predicted by anyone else were ever found.

Of all the people who set out to find a planet beyond Neptune, the most persistent was the American astronomer Percival Lowell. After founding his own observatory in Flagstaff, Arizona, Lowell spent many years observing the planet Mars and drawing maps of what he believed to be an elaborate system of canals on its surface. Lowell went on to speculate about the beings who had constructed these canals, and promoted his theories about Mars in a series of books and lectures. Although he was best known for his unconventional opinions about Mars, Lowell also harbored another, less public obsession: Like many astronomers before him, he wanted to discover a new planet.

Lowell's first search for a planet beyond Neptune took place between 1905 and 1907, when he undertook a photographic search of the zodiac. Using a magnifying glass to examine images of the sky taken a few days apart, he hoped to find a new planet, which he referred to as "Planet X." But he found nothing. So Lowell turned to mathematical analysis and derived what he believed to be the position of the new planet from gravitational perturbations. He concluded that the planet must lie in the constellation Libra, and in 1911 a new photographic search was begun, only to be abandoned the following year.

In the years that followed, Lowell continued to work on his calculations and produced several more predictions. Between 1914 and 1916, nearly 1,000 photographs were taken as part of a frantic planet search. Each time he made a new prediction, Lowell was convinced that the planet would be found almost immediately, just as Galle had found Neptune on the first night of searching. Yet no planet was ever found, and Lowell suffered a stroke and died in 1916, disappointed and exhausted.

Other members of his family were, however, determined to

prove that he had been justified in his belief that there was another planet beyond Neptune. The search was held up for several years by legal wrangling when Lowell's widow, Constance, contested her husband's will. Eventually Lowell's nephew Roger Lowell Putnam took over his uncle's observatory in 1927 and ordered the construction of a new telescope and camera specifically to carry out a new search for Planet X. By the spring of 1929 the telescope was ready, and a new assistant, a twenty-four-year-old student astronomer called Clyde Tombaugh, had been recruited to operate it.

Tombaugh began taking pictures of the zodiac, which were scrutinized for signs of the planet using a machine called a blink comparator. Comparing two photographs of a particular region of the sky for discrepancies is extremely difficult, because each photograph may contain hundreds or even thousands of individual stars. A blink comparator, which is a sort of glorified slide viewer, makes such comparisons easier by alternately displaying one photograph and then the other in its eyepiece. To a viewer looking through the eyepiece, the stars remain fixed in place as the comparator switches between the two images, but moving bodies such as planets or asteroids appear to jump back and forth. From the size of the jump, the body's distance from the Sun can be calculated.

After two months of hasty but unproductive searching, Tombaugh decided to put the search on a more organized footing. He drew up a schedule for systematically photographing a different constellation of the zodiac every month. Each region would be photographed three times, so that any suspect objects detected by comparing the first two images could be checked using the third. In September 1929, once the skies had cleared after the summer rains, Tombaugh began a new search for Planet X.

By January 1930, he had reached the constellation of Gemini. On January 21, 23, and 29, he photographed the region near the star Delta Geminorum, and on the afternoon of February 18, while comparing the second two plates, he noticed a tiny dot jumping back and forth as the blink comparator switched from one image to the other. The distance moved by the jumping dot suggested that it was the image of an object orbiting the Sun beyond Neptune. Tombaugh rushed to the office of the observatory director and announced: "I have found your Planet X."

The new object was observed and rephotographed in order to check Tombaugh's claim. He had certainly found something; but it was much fainter than Lowell had predicted, and no disk could be seen, even through the observatory's most powerful telescopes. Nonetheless, the body's position corresponded roughly (to within 6°) with one of Lowell's many predictions of the position of Planet X. So, on March 13, 1930, the seventy-fifth anniversary of Lowell's birth, the discovery of a new planet was announced to the world in a terse but triumphant telegram: "SYSTEMATIC SEARCH BEGUN YEARS AGO SUPPLEMENTING LOWELL'S INVESTIGA-TIONS FOR TRANS-NEPTUNIAN PLANET HAS REVEALED OBJECT WHICH SINCE SEVEN WEEKS HAS IN RATE OF MOTION AND PATH CONSISTENTLY CONFORMED TO TRANS-NEPTUNIAN BODY AT AP-PROXIMATE DISTANCE HE ASSIGNED." The telegram went on to give the object's position and pointed out that it agreed with Lowell's predicted longitude. It seemed that Planet X had been found, and the method pioneered by Adams and Le Verrier had, almost a century later, notched up a second success.

Tombaugh's discovery was front-page news all over the world, and attention quickly turned to the question of naming the new planet. Hundreds of names were suggested, both traditional and modern; just as William Herschel had argued after the discovery of Uranus, some people suggested that the old-fashioned tradition of mythological names should be abandoned. For her part, Constance Lowell initially suggested calling the planet Lowell, and then, even more presumptuously, insisted that the planet be named Constance—despite the fact that she had impeded the search for many years. After a while, however, two names emerged as the most popular: Minerva and Pluto. Minerva had been suggested as a possible name for Uranus and had later been put forward as a compromise solution by John Herschel during the controversy over the naming of Neptune. But since an asteroid found in 1867 had been named Minerva, the name was now taken. In any case, Pluto, the brother of Jupiter and Neptune and the god of the underworld, seemed more appropriate; its first two letters commemorated Percival Lowell, who had been instrumental in its discovery. So Pluto it was.

Yet the more astronomers observed the new planet, the stranger it seemed. For a start, its orbit crosses that of Neptune and is tilted with respect to the orbits of the other planets, rather like that of an asteroid. And although Pluto's mass was originally estimated to be seven times that of Earth, astronomers soon concluded that Pluto is in fact far smaller and less massive. As more became known about this distant, icy body, its estimated mass shrank repeatedly: By 1955, Pluto's mass was thought to be the same as Earth's; by 1968, the estimate had fallen to $\frac{1}{5}$ of Earth's mass; and eventually that fraction was reduced to $\frac{1}{500}$. Pluto's diameter is now thought to be only 1,400 miles across. Indeed, it is only just larger than Ceres, the largest member of the asteroid belt. This means Pluto is far too small and insignificant to have had any noticeable effect on the orbits of Uranus and Neptune, and could not possibly have been detected through the analysis of gravitational perturbations. The fact that it was found close to one of Lowell's predicted positions for Planet X was, in other words, simply a coincidence.

Indeed, it has been suggested that Pluto does not really deserve to be regarded as a planet at all. Given that its tenuous atmosphere freezes onto its surface when Pluto is far from the Sun, only to vaporize as it moves back toward the Sun, Pluto has been described by some mischievous astronomers as a giant comet. Others have suggested that Pluto really ought to be classified as an asteroid. Since it is currently considered to be the ninth planet of the solar system, this would give Pluto a sort of astronomical dual citizenship. A third group of astronomers contends that Pluto should be stripped of its planetary status altogether and reclassified as a "trans-Neptunian object" (TNO).

This last suggestion makes a lot of sense, and there is a good historical precedent for it. Since 1992, over 200 small, icy, Pluto-like bodies have been discovered in similar orbits beyond, or crossing, that of Neptune. This suggests that, like Ceres, Pluto is merely the largest member (and the first to be discovered) of a belt of small objects in similar orbits. So, just as Ceres was originally considered to be a planet after its discovery in 1801 but was then reclassified as the largest asteroid, it would seem to make sense to regard Pluto as the largest TNO, rather than as a planet. (The fact that Pluto has a moon, Charon, discovered in 1978, does not give it any more of a right to be considered a planet; a couple of asteroids are known to have moons, too.)

In the years after Pluto's discovery it became clear that it was not responsible for the discrepancies in the orbits of Uranus and

Neptune, so new searches for Lowell's Planet X were carried out. During the 1940s and 1950s the positions of several more planets were predicted by various astronomers in orbits 65, 75, 77, and 78 AU from the Sun. Tombaugh, who eventually stopped searching for Planet X, pointed out that his photographic searches would have been able to detect a Jupiter-like planet as far as 470 AU from the Sun, and a Neptune-like planet at a distance of 210 AU. Planet X, if it existed, was doing a very good job of concealing itself in the far reaches of the solar system.

By this time, however, some astronomers had turned their attention toward a vast new frontier: They began to search for worlds beyond the solar system, in orbit around stars other than the Sun. Evidence that other stars have planetary systems of their own would enable astronomers to begin to determine whether planets are rare or abundant in the universe, and to fathom the laws governing planetary formation. To find a planet in orbit around another star would, in other words, have far greater significance than the detection of a new world within the solar system. And once again, the method used to discover Neptune would show the way.

The idea that there might be planets in orbit around other stars goes back thousands of years. In the fourth century B.C.E. the Greek philosopher Epicurus proposed the notion of a "plurality of worlds." The universe, he suggested, consisted of innumerable spheres packed together, each containing a separate solar system. "There are," he declared, "infinite worlds both like and unlike this world of ours." But how could astronomers detect them?

Christian Huygens, a Dutch scientist and astronomer, was one of the first to consider the problem of observing planets around other stars. In his book *Cosmotheoros* (known as *Celestial Worlds Discover'd* in English), published in 1698, he wrote that he considered it unlikely that there was life elsewhere in the solar system. But, he added, that did not preclude the possibility of life on planets around other stars. Earth-like qualities, he wrote, "we must likewise grant to all those planets that surround that prodigious number of suns." But it was impossible to imagine how such planets could ever be observed.

A planet emits no light of its own but is illuminated by the star it orbits. So the feeble light reflected by a planet circling another star is drowned out by the much brighter light of the star itself when viewed from Earth. Trying to observe such a planet through a telescope is like trying to see a gnat flying around a floodlight from several miles away—only much more difficult. William Herschel, who speculated that every star "is probably of as much consequence to a system of planets, satellites and comets as our own sun," lamented that such planets "can never be perceived by us."

Some astronomers decided to have a look anyway. One of them was Hermann Goldschmidt, a German painter and astronomer who also discovered fourteen asteroids. In 1863 Goldschmidt claimed to have seen five bodies in orbit around the star Sirius. But this was the result of either wishful thinking or an optical flaw in his telescope, because nobody else could see them. Similarly, Thomas Jefferson See, the maverick American astronomer who searched for planets beyond Neptune, claimed to have seen planets associated with a star called 70 Ophiuchi. But other astronomers pointed out that any such planets, even if they did exist, would be far too faint to be visible from Earth.

The discovery of Neptune, however, suggested a different way to look for planets around other stars. Rather than searching for them directly, using a telescope, astronomers could look for indirect evidence of their existence. Just as Neptune had revealed itself through its gravitational effect on the motion of Uranus, it might be possible to detect an unseen planet from its gravitational effect on its parent star. Detecting a planet orbiting another star would be extremely difficult. But since it would resolve centuries of speculation, it would be a discovery of the utmost scientific and philosophical significance.

Since stars are so much larger and more massive than planets, it is hard to imagine that planets can affect their parent stars at all. But they can. The planet Jupiter, for example, even though its mass is only about one-thousandth that of the Sun, induces a tiny wobble in the star's motion. Strictly speaking, Jupiter does not orbit the Sun; instead, both the Sun and Jupiter orbit their common center of mass. If the Sun and Jupiter were equally massive, this point would be halfway along the line between their centers. But since the Sun is 1,000 times more massive than Jupiter, the center of mass is 1,000 times closer to the center of the Sun than it is to the center of Jupiter. (The situation is similar to a seesaw with a heavy person at one end and a lighter person at the other; the fulcrum must be nearer the heavier person for the seesaw to balance.) Both Jupiter and the Sun take 12 years to circle this point, and this causes the Sun to wobble gently from side to side by a distance roughly equal to its own radius. Observed from a vantage point 100 million million kilometers away (from a hypothetical planet in orbit around the nearby star Procyon, for example), the Sun would appear to wobble from side to side by 1.6 thousandths of an arcsecond with every 12-year orbit of Jupiter.

This is an absolutely tiny amount—less than a thousandth of the size of the discrepancies in the position of Uranus that led to the discovery of Neptune. Seen from twice as far away, the Sun's wobble would be half as big, and at greater distances it would be smaller still. But the advent of astro-photography has made the determination of stellar positions—a practice known as astrometry—far more accurate than it was in the days when astronomers relied on telescopes and quadrants. By repeatedly photographing the region of the sky around a star, an astronomer can determine its position in relation to its neighbors. It should then be possible to see whether or not the star is wobbling—and, therefore, whether or not it has planets.

With so many stars in the sky, where should an astronomer begin searching for planets by astrometry? The answer is simple: The closer a star is to our solar system, the easier it will be to detect any wobble. By far the most famous astrometric planet search was that carried out by Peter van de Kamp, who spent decades observing a small, faint star called Barnard's Star.

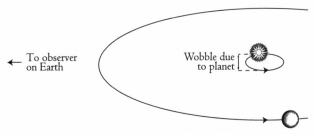

The side-to-side wobble of a star caused by an orbiting planet, greatly exaggerated.

This star, which was discovered in 1916 by the American astronomer Edward Emerson Barnard, is the third-closest star to
the Sun. It is 56 million million kilometers away, a small distance
by astronomical standards. Light from Barnard's Star takes 6
years to reach the Earth, so the distance to Barnard's Star is more
conveniently expressed as 6 light-years. As well as being nearby,
Barnard's Star is less massive than the Sun, which means that a
Jupiter-like planet in orbit around it would cause a much larger
astrometric wobble. Detecting such a planet would, however, require several years of observation.

Van de Kamp started photographing the region around Barnard's Star in 1938 from Sproul Observatory in Pennsylvania,
using a powerful 24-inch telescope. By the early 1960s he believed
he had found evidence that Barnard's Star was indeed wobbling—by a massive 24.5 thousandths of an arcsecond. The astrometric wobble, he declared with a confidence worthy of Le
Verrier, "admits of no simpler explanation than that of a perturbation caused by an unseen companion of Barnard's Star." From
the size of the wobble, he could determine the companion's
mass, and from the period of the wobble, he could determine
how long the companion took to orbit the star. Van de Kamp
concluded that Barnard's Star was orbited by an unseen planet
with almost twice the mass of Jupiter that took 24 years to complete each orbit. Unusually, the planet appeared to have a long,
elliptical orbit, rather like a comet's, so that its distance from its
parent star varied between 2 and 7 AU. The technique of astrometry had, it seemed, revealed its first planet.

By 1969 van de Kamp had refined his prediction. Based on
thousands of observations, he now believed there were two giant
planets in roughly circular orbits around the star, with periods of

12 and 26 years, respectively. This was an exciting finding, because it suggested that the arrangement of planets around Barnard's Star was similar to that of the planets around the Sun. (The periods of Jupiter and Saturn are 12 and 30 years, respectively.) This in turn suggested that planetary systems like the Sun's were commonplace, and that Barnard's Star might also have smaller, Earth-like planets. It would not, however, be possible to detect any such planets astrometrically because they would not be massive enough to induce a detectable wobble in their parent star.

Not all astronomers were convinced by van de Kamp's predictions, however. Other observers who measured the position of Barnard's Star failed to detect any wobble at all. A detailed study published in 1973 by George Gatewood and Heinrich Eichhorn analyzed the position of Barnard's Star with painstaking precision and found no evidence whatsoever of any planets. Subsequent analysis of van de Kamp's data suggested that the wobble he had detected had actually been caused by a realignment of his telescope during cleaning in 1949. Even after discarding his pre-1950 observations, however, van de Kamp continued to insist that he had solid evidence that his two planets really existed.

But the idea of detecting planets by astrometry was quickly gaining a dubious reputation. In addition to van de Kamp's claims about Barnard's Star, various astronomers claimed to have found planets around the nearby stars 70 Ophiuchi, 61 Cygni, Lalande 21185, and Epsilon Eridani, using similar methods. Yet none of these claims stood up to scrutiny by other observers, which suggested that they were all the result of instrumental errors.

Interest in Planet X was revived in the 1970s and 1980s following the dispatch of the unmanned Pioneer and Voyager probes to the outer solar system. Several new predictions were made of Planet X's size and position, based once again on analysis of apparent anomalies in the motions of Uranus and Neptune.

Ultimately, however, it was the planet Neptune that demolished the case for Planet X. In 1989, the Voyager 2 probe passed by Neptune, and from the change in the spacecraft's trajectory it was possible to work out Neptune's mass with unprecedented accuracy. It turned out that Neptune was 0.5 percent less massive than had been previously thought. E. Myles Standish, an astrophysicist at NASA's Jet Propulsion Laboratory, inserted this more accurate value for Neptune's mass into the orbital models for the outer planets and found that the remaining anomalies in the orbits of Uranus and Neptune—which had formed the basis for a century of speculation about planets beyond Neptune—almost completely vanished. His results were widely seen as the final nail in the coffin of Planet X.

Images sent back by Voyager 2 also revealed the nature and extent of Neptune's rings, whose existence had been suspected since the mid-1980s on the basis of ground-based observations. The rings, some of which are strangely incomplete and form partial "arcs," were subsequently named in honor of the astronomers involved in the planet's discovery. As a result Adams, Le Verrier, Galle, and Arago all have rings named after them—though Airy and Challis do not. A ring arc was also named after William Lassell, the British astronomer who discovered Neptune's moon Triton.

By rekindling public interest in the distant planet, the Voyager flyby also had the effect of resurrecting the Neptune controversy, which had largely died down during the twentieth century, except for a brief outburst in 1946 on the occasion of the centenary of the discovery. On that occasion W. M. Smart, of Glasgow University, was asked to give a commemorative address to the Royal Astronomical Society, in which he retold the story of the Neptune affair. He was subsequently criticized by the astronomer royal of the day, H. Spencer Jones, who accused him of being unjust in his criticism of Airy, and the two astronomers exchanged argumentative letters in the pages of the journal *Nature.*

Things then quieted down again until the late 1980s, when another flurry of debate began. Allan Chapman, a historian at Oxford University, published a defense of Airy; Robert W. Smith of the Smithsonian Institution highlighted the significance of the Cambridge connection in the English search for Neptune; and a biography of Adams, written by his great-great-niece Hilda Harrison, resurrected the idea that Airy had secretly conspired with Le Verrier. And when it emerged that Airy's "Neptune File"—his scrapbook containing many of the original documents relating to the discovery—had mysteriously disappeared, another historian, Dennis Rawlins, accused the Royal Greenwich Observatory of a cover-up and of trying to suppress its contents.

Almost a century and a half after the discovery of Neptune, it seemed that nothing had changed. The arguments surrounding the discovery continued—and the promise of using the law of gravitation to detect new worlds without seeing them remained unfulfilled. Despite numerous attempts, no new planets had been found, either within the solar system or beyond it. Decades of theorizing and searching had revealed only Pluto, a tiny orbiting snowball whose right to be regarded as a planet at all seems increasingly dubious, and whose discovery was the result of a systematic trawl of the heavens, rather than the observational confirmation of a theoretical prediction. Neptune remained in a class by itself, the controversy of its discovery unresolved, the triumph of its detection unsurpassed, and the legacy of its discoverers unclaimed.

Worlds Unseen

Observe how system into system runs,
What other planets circle other suns.

—ALEXANDER POPE
AN ESSAY ON MAN

By the early 1990s planet hunting was seen as a somewhat disreputable field of astronomy. Funding to search for planets around other stars was extremely difficult to come by; astronomers involved in the field were regarded with suspicion. But still they searched.

The idea of using astrometry to detect planets by measuring the wobbles of their parent stars, which had originally seemed so promising, had failed completely. So planet hunters turned to a different method of indirect detection, called the "radial velocity" technique. Like astrometry, it involves searching for the wobble of a star caused by the gravitational influence of an orbiting planet. But unlike astrometry, which measures the side-to-side wobble of a star, the radial velocity technique measures the star's back-and-forth motion along the line of sight between the star and the observer.

For example, viewed from a hypothetical planet in orbit around the nearby star Procyon—about 100 million million kilometers (11 light-years) away—Jupiter causes the Sun to move from side to side by 1.6 thousandths of an arcsecond with each 12-year orbit, as both planet and star revolve around their common center of mass. But Jupiter also causes the Sun to move back and forth relative to an observer on our hypothetical planet. This motion causes tiny changes in the Sun's spectrum—the signature of light and dark lines that is produced when its light is shone through a prism. The changes are due to a phenomenon called a Doppler shift. This is familiar in everyday life as the effect that causes the siren of an approaching ambulance to sound higher pitched as the ambulance approaches and lower pitched as it recedes, because the sound waves get bunched up in front of the moving ambulance and spread out behind it. A similar effect happens with a moving star, causing the lines in its spectrum to shift one way as the star approaches an observer and the other way as it recedes.

From the size of this shift it is possible to determine the speed the star is moving along the line of sight to the observer; this speed is called the "radial velocity." Jupiter, for example, causes the Sun to move back and forth with a maximum radial velocity of 12.5 meters per second. So a distant observer on a planet

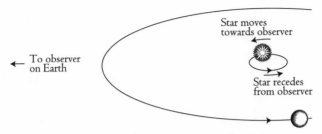

The back-and-forth wobble of a star caused by an orbiting planet, greatly exaggerated. Repeatedly measuring the star's speed along the line of sight to the observer (the radial velocity) makes it possible to detect the presence of the planet.

around Procyon would be able to detect Jupiter by repeatedly examining the Sun's spectrum over a period of several years, measuring the shift of its spectral lines, determining its radial velocity, and then looking for a periodic variation. Similarly, observers on Earth ought to be able to detect planets in orbit around other stars by looking for radial velocity wobbles.

Just as with astrometry, in the event of a wobble being found, the mass of the planet causing it can be estimated from the wobble's size, and the time taken by the planet to orbit its parent star can be determined from the wobble's period. The larger the planet, and the closer it is to its parent star, the bigger the wobble.

Complicated though it sounds, the radial velocity technique has an important advantage over astrometry: The back-and-forth motion of a star caused by an orbiting planet does not diminish when viewed from a greater distance. (By contrast, the farther away you are from a star, the smaller a side-to-side astrometric wobble appears, making it harder to see.) This means that planet hunters using the radial velocity technique have a far larger number of target stars to choose from; they need not limit themselves to the handful of stars in the Sun's immediate neighborhood.

The first planet search using the radial velocity method was carried out by two Canadian astronomers, Bruce Campbell and Gordon Walker, starting in 1980. Their approach was to point their telescope toward a Sun-like star and direct its light through a gas cell containing hydrogen fluoride, which superimposes its own characteristic (but fixed) pattern of light and dark lines over the star's spectrum. By using the hydrogen fluoride's spectrum as a yardstick, they were then able to measure the positions of the star's spectral lines with sufficient accuracy to determine the star's radial velocity to within 15 meters per second. This degree of accuracy is not sensitive enough to detect a planet of Jupiter's size and orbital distance, because a Jupiter-like planet causes only a 12.5-meter-per-second radial velocity wobble in a Sun-like star. But it is capable of picking up heavier planets or planets closer to their parent stars than Jupiter is to the Sun (either of which would cause the parent star to move back and forth more quickly than 12.5 meters per second).

Campbell and Walker observed twenty-one stars for nearly fifteen years. There was a false alarm in 1988, when they claimed to have found tentative evidence for several planets. Ultimately their search was fruitless, and they abandoned it in 1995. By this time, however, several other groups of astronomers had also started looking for planets using the radial velocity method.

Geoffrey Marcy, an astronomer at San Francisco State University, began his search in 1987. He and his colleague Paul Butler decided to use iodine instead of hydrogen fluoride in their gas cell in order to improve the accuracy of their measurements. Iodine has more dark bands in the relevant part of the spectrum, which makes it a more precise yardstick. But keeping track of the extra dark bands makes the process of calculating the radial velocity much more complicated, so that it can only be accomplished using powerful computers. After years of refining their technique, by 1994 Marcy and Butler could measure the radial velocity of a star to within 3 meters per second.

Meanwhile, another group led by William Cochran at the University of Texas at Austin had launched a similar planet search using the radial velocity technique. And two Swiss astronomers, Michel Mayor and Didier Queloz, began a search of their own from the Haute-Provence Observatory in the south of France. The Swiss astronomers, however, had devised a tech-

Geoffrey Marcy (left) and Paul Butler (right) (Keck Observatory)

nique with a clever twist: They did away with the gas cell. Instead
of passing the light from the target star through such a cell, and
thus mixing up the star's spectrum with the yardstick provided
by the gas, Mayor and Queloz simultaneously (but separately)
observed both the star and a lamp whose spectral signature was
precisely known. This meant that the star's spectrum and the
yardstick provided by the lamp did not have to be disentangled.
Queloz wrote software that could extract a star's radial velocity
within minutes of an observation, and to within 13 meters per
second. So unlike Marcy and Butler, whose technique required
days of number crunching to determine the radial velocity values,
Mayor and Queloz could work out the radial velocity for each
observation of each star as they went along. The two astronomers
began examining 142 Sun-like stars for evidence of planets in
1994, observing each star every couple of months. Like everyone
else in the field, they assumed that several years of observations
would be required to reveal trends in the stars' radial velocities
caused by orbiting planets.

In October 1994, however, Queloz noticed something odd
about the star 51 Pegasi, a Sun-like star in the constellation of
Pegasus, the winged horse. The first few observations of the star
had produced similar values for the radial velocity, but the October observation produced a totally different value. When the December observation also yielded a wildly different value, Queloz
initially concluded that there had to be something wrong with
his software. He and Mayor decided to observe the star more
closely during December, and by January 1995 it was clear that
the software was not at fault; 51 Pegasi was shaking back and
forth at an extraordinary rate. Rather than taking several years to
complete each wobble, as would be expected of a star orbited by

Didier Queloz (left) and Michel Mayor (right) (Geneva Observatory)

a Jupiter-like planet, 51 Pegasi seemed to be wobbling far more quickly. Indeed, Queloz calculated that it was completing one wobble every four days, a result that was confirmed by further observations completed at the beginning of March. Queloz sent Mayor, who was abroad on sabbatical, a fax in which he suggested that 51 Pegasi's motion might be due to an orbiting planet. If so, it was unlike any planet in the solar system: a huge, Jupiter-like world, but orbiting even closer to its star than Mercury orbits the Sun.

Mayor and Queloz spent the next few weeks trying to convince themselves that they really did have a planet on their hands. They were all too aware of the numerous false alarms that had dogged the field of planet hunting, so they were reluctant to jump to conclusions. At the same time, they were terrified that another of the teams of astronomers would also notice 51 Pegasi's huge variation in radial velocity—the star was moving back and forth by as much as 60 meters per second—and would beat them to the discovery. They considered whether the variation in radial velocity might be due to a bulge or spot on the star itself, or some kind of pulsation of its surface. But none of these explanations stood up to scrutiny. With growing excitement, they checked their instruments and their software for faults, but found none. No other explanation made sense; 51 Pegasi had to have a planet.

Before announcing their findings, Mayor and Queloz set themselves a final test. They wrote up their results for publication and drew up a chart predicting the radial velocity of the star for four days in the first week of July. Their plan was to observe the star on four successive nights, to ensure that its radial velocity matched their predictions. If it did, they would then submit their paper to the scientific journal *Nature* for publication.

On the first night, the two astronomers anxiously pointed their telescope at 51 Pegasi and waited for the computer to spit out the value for its radial velocity. It matched their prediction perfectly. On the following two nights they repeated the procedure, and in both cases the radial velocity accorded with their predictions. On the fourth night, the two men took their families with them to the observatory to make the final, decisive measurement. When it too came out as expected, they brought out a cake and opened a bottle of champagne to celebrate.

Mayor and Queloz announced their discovery at an astro-

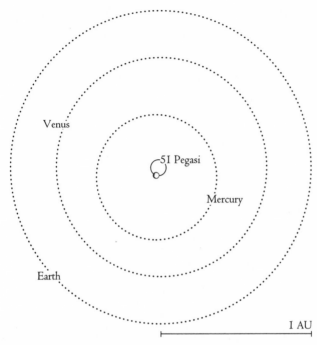

The orbit of the planet around 51 Pegasi, with the orbits of Mercury, Venus, and Earth shown on the same scale for comparison. 51 Pegasi's planet is a gas giant like Jupiter, yet it orbits much closer to its parent star than Mercury orbits the Sun. For clarity, the central star is not shown.

nomical conference in Florence, Italy, on October 6, 1995. Their mathematical analysis of their radial velocity observations showed that there had to be an unseen planet in orbit around 51 Pegasi. Their results suggested that the planet was comparable in mass to Jupiter and orbited the star once every 4.2 days at a distance of 0.05 AU. This was, to be sure, an extremely unusual planet. In defiance of the accepted theories of planetary formation, which suggested that Jupiter-like planets could form only more than 4 AU from their parent stars, this new planet had somehow ended up in a much closer, tighter orbit. But that was a question for the theorists to worry about; the important thing was that, for the first time, a planet had been detected orbiting another sun. It would be the first of many.

Confirmation of Mayor and Queloz's results came swiftly. Marcy and Butler, whose list of target stars had not included 51 Pegasi, started taking measurements of its radial velocity a few days later and found that it was indeed wobbling in just the way that Mayor and Queloz said it was. Within days, they had gathered enough evidence of their own to confirm the presence of a planet. Their announcement that they had verified the planet's existence caused a worldwide sensation. At last, two independent teams of planet hunters had pointed their telescopes at the same star and produced positive results that agreed with each other perfectly—something that had never happened in the bad old days of hunting for Planet X or looking for planets using astrometry.

Marcy and Butler realized that, having collected almost eight years of radial velocity measurements, they might also have found other planets without knowing it. They had assumed, along with the rest of the astronomical community, that the sort of large, Jupiter-like planets they were looking for would take a decade or so to orbit their parent stars. As a result, it would not be worth analyzing their observations until they had monitored their target stars for several years. But 51 Pegasi had proved everyone wrong. Because it was so close to its parent star, and completed its orbit quickly, it could be detected after just a few days of observing. This meant that if any of the stars on Marcy and Butler's target list had planets in such close, tight orbits, the astronomers would already have collected enough data to detect them. It was simply a matter of analyzing their observations.

In November 1995, the two astronomers started sifting through their mountain of data—a time-consuming process that required vast amounts of computer power. The following month their analysis turned up evidence of a planet orbiting the star 47 Ursae Majoris, a Sun-like star in the constellation of the Great Bear (part of which is better known as the Big Dipper). Subsequent analysis revealed that the planet was about three times as massive as Jupiter and orbited its star at a distance of 2.1 AU, completing one orbit every three years or so. Evidence of the planet's existence had been sitting in Marcy and Butler's computers all along, right under their noses.

On December 30, a second discovery emerged from the data: evidence for a body roughly eight times as massive as Jupiter orbiting the star 70 Virginis in a highly elliptical orbit. Marcy and Butler announced these two discoveries in January 1996. Now that the world's planet hunters knew what they were looking for, further discoveries soon followed. Within months, new planets had been detected around the stars Tau Bootis A, Rho Cancri A, Upsilon Andromedae, Rho Corona Borealis, and 16 Cygni B. After that the discoveries came thick and fast, as new teams of astronomers joined the hunt. In addition to the eight planets found between 1995 and 1997, eight more planets were found during 1998, and another twelve by the end of 1999, by which time a new planet was being found, on average, every month. At the time of writing (May 2000), the total had reached forty-one.

As the list of planets continues to grow, astronomers and planetary theorists have become increasingly puzzled. The planet around 51 Pegasi turned out to be just the first of several giant planets that have been found orbiting very close to their parent stars. Similarly, several planets have been found that, like the planet around 70 Virginis, have curiously elongated orbits. This means that the accepted theories of planetary formation, which cannot account for either type of planet, will need to be rewritten. The idea that planetary systems around other stars will be broadly similar to our own solar system is no longer tenable. Indeed, as more planets are discovered, it is our own solar system itself that starts to seem more and more unusual.

The unexpected nature of these extrasolar planets has led some skeptics to suggest that the planets do not really exist. After Mayor and Queloz published their evidence that 51 Pegasi was wobbling, Canadian astronomer David Gray claimed that according to his observations the wobble was actually due to complex pulsations within the star itself, rather than the gravitational influence of a planet. Mayor and Queloz collaborated with Marcy and Butler to write a detailed refutation, and when subsequent observations of 51 Pegasi failed to agree with the pulsation hypothesis, Gray retracted his claim.

Other skeptics have suggested that the extrasolar planets that have been detected are not planets at all but very small, dim stars called brown dwarfs. In fact, the distinction between a very large planet and a very small star is not at all clear. Both are, essentially, big balls of hydrogen and helium held together by gravitational attraction. As yet, astronomers are uncertain if there is an upper limit on the mass of a planet or a lower limit on the mass of a star. The planet orbiting 70 Virginis, for example, at around eight Jupiter masses, is only slightly less massive than the smallest known brown dwarf.

Two results, however, have provided dramatic support for the idea that the bodies detected in orbit around distant stars really are planets, rather than dwarf stars. In April 1999, the first evidence was announced of multiple planets around a Sun-like star. The star in question, Upsilon Andromedae, was already known to have a Jupiter-mass planet orbiting it every 4.6 days at a distance of 0.06 AU. Marcy and Butler teamed up with other planet hunters—at the Harvard-Smithsonian Center for Astrophysics in Cambridge, Massachusetts, and the High Altitude Observatory in Boulder, Colorado—and analyzed their data back to 1988. From the complex wobbling of the star, they found evidence for another two planets, of two and four Jupiter masses, orbiting Upsilon Andromedae at 0.83 AU and 2.5 AU, respectively. The three superimposed wobbles, each caused by a different planet, were separated using computer analysis.

The fact that Upsilon Andromedae has three planets is significant because stars and planets are thought to form in different ways. A star forms via the gravitational collapse of a cloud of dust and gas, whereas planets form out of the disk of leftover material swirling around a young star. Only the second process can result in several small bodies orbiting one much larger one. So the presence of multiple bodies in concentric orbits around Upsilon Andromedae strongly suggests that they really are planets. Other multiple-planet systems are likely to be detected in the coming years as astronomers continue to observe stars around which single planets have been detected and look for longer wobbles superimposed on their motion. Several stars that are known to have at least one planet are already suspected of having another.

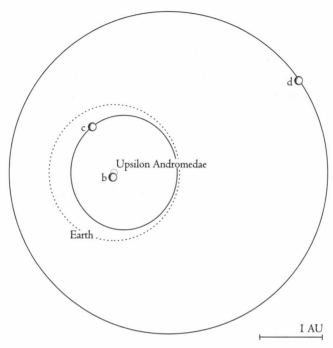

The orbits of the three known planets in the Upsilon Andromedae system (b, c, and d), with the orbit of the Earth shown on the same scale for comparison. For clarity, the central star is not shown.

Further confirmation of the planetary nature of the bodies that have been detected orbiting other stars came in November 1999. Many astronomers began monitoring the brightness of stars known to have planets, in the hope that one of these planets would pass directly between its star and the Earth and cause a slight dimming of the star—in other words, a transit. This can only happen if the planet's orbital plane is aligned so that it is viewed nearly edge-on from Earth, something that statistically ought to be the case for about one in ten such planets. In the summer of 1999, two independent groups of planet hunters detected a Jupiter-mass planet orbiting a star called HD 209458 at a distance of 0.05 AU with an orbital period of 3.523 days. The planet's orbit turned out to be aligned edge-on when viewed from Earth, and the star was seen to dim by about 2 percent on several separate occasions, by several independent groups of observers, at exactly the moment when the planet was predicted to pass in front of it.

This means that during the transit, the unseen planet obscured 2 percent of the star's disk as seen from Earth. This observation made it possible to determine the size and mass of the planet very precisely. It was found to be 63 percent as massive as Jupiter, and about 1.5 times bigger—in other words, it is a gas giant whose proximity to its parent star has prevented it from cooling and shrinking as it would if it were in a colder, more distant orbit. Furthermore, its size is exactly that predicted by a new theoretical model of "blowtorched" planets orbiting close to their parent stars, which shows that the new theories being devised to explain the origins and properties of unusual extrasolar planets are on the right track. But most important, the observations of the transit of an extrasolar planet provide striking evidence that such planets really do exist, and that the wobbles detected by the radial velocity technique are not caused by brown dwarfs, stellar spots, bulges, or pulsations.

Now that it is certain that they are real, the question has inevitably arisen of how best to name extrasolar planets. At the moment, the planets are referred to by appending a small letter onto the name of their parent stars. The planet around 51 Pegasi, for example, is called 51 Pegasi b, and the three planets around Upsilon Andromedae are known as Upsilon Andromedae b, c, and d, in order of their distance from the parent star. (This convention is an extension of the scheme used to name binary stars, so the letter "a" is technically assigned to the star itself.)

But there are a number of drawbacks with this scheme. If a star is known to have one planet (b), and a second planet is subsequently discovered in a closer orbit to the star, the first planet has to be renamed from b to c, which could cause problems: Any scientific papers, for example, that previously referred to the first planet as b would then be using the wrong name. This scenario is more than possible; it is extremely likely. Current methods are unable to detect small planets the size of Earth or Mars around other stars, but once it is possible to detect such planets, they may well be found closer to their parent stars than already-known giant planets.

Of course, an easier solution might be to assign names to the new planets, similar to the names given to planets and moons within our own solar system. Inevitably, this raises the question of tradition: Should the new planets have mythological names? Traditionalists have proposed the mythological name Bellerophon for 51 Pegasi's planet; in Greek mythology, Bellerophon was the hero who tamed Pegasus, the wild horse. For their part, Michel Mayor and Didier Queloz have suggested the name Epicurus for the planet, in honor of the philosopher who first suggested the possibility of the existence of other solar systems. They argue, only half seriously, that a new kind of naming convention will be necessary to keep up with the flood of discoveries, otherwise the stock of mythological names will soon run out; so they propose naming extrasolar planets after real historical figures, rather than fictional characters.

So far, however, the International Astronomical Union, which has the authority to name astronomical objects, has not adopted a naming scheme for extrasolar planets. But given the rate at which such planets are now being discovered, it will soon be time to draw one up. At that point, perhaps the mythological tradition will finally be discarded.

Since 1995 planet hunters have triumphantly secured a clutch of planets, and they will detect dozens of new worlds in the coming years. They are, however, all too aware of the way the radial velocity technique skews their understanding of planetary systems. The surface of a star is not smooth and static but is constantly and randomly bubbling and moving around. For relatively stable stars, the surface moves at around a meter per second, and for less stable stars the surface moves more quickly. This random movement can be separated from the regular, smooth wobble caused by an orbiting planet, provided the wobble is big enough. But the tiny wobbles caused by planets that are small, or far away from their parent star, get drowned out by the activity on the star's surface. As a result, the radial velocity technique can detect only really massive planets and those orbiting very close to their parent stars.

An alien astronomer on a distant planet measuring the radial velocity of the Sun would initially conclude that the Sun has only one planet, Jupiter, which causes a wobble of 12.5 meters per second. After several decades of observing, it would just be possible to detect Saturn, which has less than a third of the mass of Jupiter and causes a wobble of about 3 meters per second. But our alien astronomer would not be able to detect Mercury,

Venus, Earth, or Mars, none of which is massive enough to cause a detectable wobble. Earth, for example, has only $^1/_{318}$ of the mass of Jupiter. Nor would it be possible to detect Uranus or Neptune, both of which are so far from the Sun that, despite their relatively large masses (about $^1/_{20}$ of that of Jupiter), they do not cause a detectable wobble either. (And as for tiny Pluto—you must be joking.)

So unless new, more sensitive methods of detecting planets are found for use alongside the radial velocity method, planet hunters will have a very distorted view of the sorts of planetary systems that exist around other stars. Fortunately, new methods have already been devised, and the next generation of planet-hunting hardware is under construction.

The method that is likely to prove most useful in the immediate future is, surprisingly, astrometry—though in an updated, more precise form, using a cunning technique called interferometry. This involves combining the light from two or more separate telescopes in such a way as to mimic a much larger telescope. If done in the right way, it is possible to arrange several small mirrors so that they behave like fragments of a single, much larger mirror. Two 10-meter telescopes positioned 100 meters apart can then be used to measure the position of a star with the same accuracy as a single 100-meter telescope. Several observatories around the world are being set up specifically to exploit this technique, including the Keck Observatory in Hawaii, the Very Large Telescope in Chile, and the Large Binocular Telescope in Arizona.

The use of interferometry to measure stellar positions will allow astrometric wobbles to be measured far more accurately than ever before and enable astronomers to detect planets from

the wobbles induced in their parent stars. Unlike the radial velocity method, which is best at picking up small, fast wobbles in a
star's motion, astrometry is best at detecting slow, gentle wobbles
caused by large planets in distant orbits from their parent stars.
The planet around 51 Pegasi, for example, causes it to wobble
back and forth once every four days. Uranus, on the other hand,
causes the Sun to wobble back and forth once every eighty-four
years. This is such a gentle wobble that the Sun's velocity hardly
changes at all—in fact, it changes by less than a meter per second—as it moves back and forth under the influence of Uranus.
So Uranus-like planets could never be detected using the radial
velocity method. But they could be detected using astrometry.
In fact, astronomers using the Keck telescopes as an interferome-
ter expect to be able to measure astrometric wobbles as small as
20 millionths of an arcsecond. This will enable them to detect
Uranus-like planets orbiting stars as far as 33 light-years from
Earth.

Even more accurate astrometry is possible from space, beyond
the distorting effects of the Earth's atmosphere, which makes
stars seem to wobble around and makes their positions harder to
measure. Unmanned space probes launched in the next few years
will be able to measure stellar positions—and stellar wobbles—
even more accurately than ground-based observatories, again
using interferometry. Space Technology 3, a probe planned for
launch in 2003, will consist of two separate spacecraft, which will
fly in formation as much as a kilometer apart in order to test
the principle of space-based interferometry using independently
maneuverable telescopes. Another planned probe, the Space Interferometry Mission (SIM), due for launch in 2005, will have
two telescopes mounted on opposite ends of a 10-meter boom

and will be able to detect astrometric wobbles as small as a mil-lionth of an arcsecond. This will enable planets as small as four Earth-masses to be detected around stars up to 33 light-years away. In theory, SIM could also detect Earth-sized planets around nearer stars.

All these new techniques will add dozens of new specimens to the menagerie of extrasolar planets. Collectively they will provide much more information about the variety of planetary systems that exist around other stars. No doubt there will be further sur-prises. Can Earth-like planets coexist in solar systems with rap-idly orbiting giant planets? Are there underlying rules that determine the size and spacing of planets in alien solar systems, as Bode's law was once thought to explain our own? (Some theorists already suspect that there are; Bode's law, it turns out, is not entirely dead after all, and similar but more sophisticated ideas are now being explored by several theorists.) These are the sorts of questions that the discoveries of the next decade will help to answer.

As a new era of planet hunting began, another ended. The triumphant discovery of the planet around 51 Pegasi, and the other extrasolar planets that have been found since, brought to an end the century and a half of fruitless planet hunting that began after the discovery of Neptune. Eventually, after almost 150 years of searching, astronomers detected a new planet with-out seeing it. The drought of planetary discoveries ended, and a deluge began. The skies are now raining planets.

We are living at the beginning of a golden age of planet hunt-ing. Hardly a month goes by without the announcement that yet another new planet has been found wheeling around a distant star. Within a decade, it will be possible to draw maps of several alien solar systems, showing the orbits and characteristics of their planets, both large and small. None of these planets, however, will have been seen by human eyes. Instead, they will have been detected indirectly, through the mathematical analysis of gravitational perturbations, just as the existence of the planet Neptune was deduced by John Couch Adams and Urbain Jean-Joseph Le Verrier in the 1840s. The discovery of Neptune is essentially the prototype of today's discoveries of planets around other stars—an astronomical dress rehearsal, within our own solar system, for the method now being used to detect countless unseen worlds.

Didier Queloz, codiscoverer of the planet around 51 Pegasi, likes to talk about Neptune when he is asked to describe his work, and in particular to explain how planets can be detected without being seen. "The idea is to try to explain to people that we are not observing these planets," he says. "We do not need to get a picture to make sure the planet is there. The way to explain that is to explain that people detected Neptune without seeing it."

The discoveries of the late 1990s have fulfilled Neptune's promise. And another discovery may finally bring the controversy surrounding the planet's discovery to a close. In the spring of 1999, Airy's lost scrapbook—the "Neptune File"—unexpectedly reappeared. It had been borrowed by a member of the staff at the Royal Greenwich Observatory, who subsequently moved to another post in Chile and took the file with him. Following his death, the file was returned to England. The reappearance of the file means that the last ghosts of the Neptune controversy—such as the idea that Airy had conspired with Le Verrier—can now be laid to rest. The file contains no evidence for this curious theory.

Even so, echoes of the Neptune story will continue to reverberate as further discoveries are made in the coming years. In particular, today we see each new extrasolar planet, to borrow the words of John Herschel's 1846 speech, "as Columbus saw America from the shores of Spain. Its movements have been felt, trembling along the far-reaching line of our analysis, with a certainty hardly inferior to that of ocular demonstration." When Herschel said these words, both Adams and Le Verrier had predicted the existence of Neptune, and Challis was already weeks into his ill-fated search. It was not until two weeks later that Johann Gottfried Galle saw Neptune in his telescope. So we are now, in a sense, halfway through a modern recapitulation of the Neptune story, as we await the first direct observation—the first pictures—of a planet around another star.

Plans for gigantic, space-faring telescopes, able to produce such pictures, are already on the drawing board. One design, proposed by the American space agency NASA for launch around 2012, is for a space-based observatory called the Terrestrial Planet Finder (TPF). Four 3.5-meter telescopes will fly in formation hundreds of meters apart. The light from the four telescopes will be beamed to a fifth craft, which will combine it in such a way that the whole system will be equivalent to a giant telescope, a kilometer across, enabling it to produce images of entire solar systems around nearby stars. A similar scheme, called Darwin, has also been proposed by the European Space Agency.

Even more ambitious is NASA's extremely tentative plan for a mission called Planet Imager. This would consist of dozens of

TPF-like craft, each with an 8-meter telescope, flying in a vast formation 6 kilometers across. Combining the light from these telescopes would make it possible to produce images of planets around nearby stars, showing their continents, weather systems, and moons. It would even be possible to examine the composition of their atmospheres for evidence of life.

Admittedly, at the moment it is hard to believe that such elaborate, expensive schemes will ever be put into operation. But Adams and Le Verrier would no doubt have been astonished to hear that, within 150 years of their discovery of Neptune, a space probe flew past the planet and beamed back pictures of it, and its rings and moons.

TPF and Darwin are not the only competitors in the race to produce the first picture of an extrasolar planet. It could also come from one of the next generation of ground-based observatories, which may be able to see large, bright, Jupiter-like planets around nearby stars, using a technique called adaptive optics to compensate for the distorting effects of Earth's atmosphere. Another possibility is that the Next-Generation Space Telescope, the planned replacement for the orbiting Hubble telescope, will produce the first image.

One way or another, chances are that within the next few years, an image of an alien planet will slowly unfurl on an astronomer's computer screen and will then go on to become front-page news all over the world. Subsequently, the first picture of an Earth-like planet in orbit around another star is likely to become an iconic image, just as the first pictures of Earth taken from space did in the 1960s.

Clearly, planet hunting is a field in which the greatest discoveries have yet to be made. But as well as encouraging us to look to the future, the success of modern planet hunters suggests a new way to look at the past. In the light of the discovery of the first extrasolar planets, Adams and Le Verrier are revealed not as the rival discoverers of Neptune but as the joint founders of the modern discipline of planet hunting. Their novel approach, pioneered in the 1840s, continues to underpin the search for new worlds today. As a result of their detective work, Uranus lit the way to Neptune—and Neptune now points the way to the stars.

Notes

1: THE MUSICIAN OF THE SPHERES

The description of William Herschel's discovery of Uranus is drawn largely from Lubbock (*The Herschel Chronicle*), Herschel's own account, which appears in the *Philosophical Transactions of the Royal Society,* and also from Sidgwick (*William Herschel*) and Armitage (*William Herschel*). Additional biographical details about William and Caroline Herschel are drawn from Clerke (*The Herschels and Modern Astronomy*), Holden (*Sir William Herschel*), and Macpherson (*Makers of Astronomy*); excerpts from Caroline's diary and the story about Mr. Bernard's music lesson are quoted from Lubbock. The determination of the planetary nature of Uranus' orbit and the debate over the planet's name are discussed by Lubbock, Arago (*Popular Astronomy*), and Moore (*William Herschel*). Matthew Turner's "terra incognito" comparison is quoted from Schaffer (*Uranus and Herschel's Astronomy*). Fans of Patrick O'Brian's Aubrey/Maturin books will note that Caroline Herschel makes an offstage appearance as a source of marital stress and astronomical inspiration in the fourth novel in the series, *The Mauritius Command.*

2: SOMETHING RATHER BETTER THAN A COMET

Further biographical details about William Herschel's later life and the construction of the forty-foot telescope are drawn from Sidgwick (*William Herschel*) and Dunkin (*Obituary Notices of Astronomers*), where the details of Herschel's sale of telescopes can also be found. (Sidgwick also describes Herschel's audience with Napoleon in 1802, at which the topics of conversation were astronomy, horse breeding, the relative merits of the Paris and London police forces, and the terrible reputation of the British tabloid press.) Details of the origin of Bode's law are drawn from Nieto (*The Titius-Bode Law of Planetary Distances*). Biographical details about von Zach and the setting up of the "celestial police" are drawn from Cunningham (*The Baron and His Celestial Police*) and von Zach's own account in

Monatliche Correspondenz. The description of Piazzi's discovery of Ceres, including his correspondence with Oriani, is drawn from Cunningham (*Giuseppe Piazzi and the Missing Planet*). Details about Gauss and his new methods are drawn from Dunnington (*Carl Friedrich Gauss: Titan of Science*), Bell (*Men of Mathematics*), and Teets and Whitehead (*The Discovery of Ceres*). The description of the lamp micrometer is from Armitage (*William Herschel*). The *Edinburgh Review*'s comments are quoted from Clerke (*The Herschels and Modern Astronomy*).

3: A VERY BADLY BEHAVED PLANET

Astronomical instruments and their use in measuring the positions of celestial objects are described by Chapman (*Dividing the Circle*). The reputation of the Royal Greenwich Observatory and the value of its unbroken sequence of planetary observations were noted by McCrea (*The Royal Greenwich Observatory*). The description of the failed attempts to determine the orbit of Uranus is drawn from the detailed account in Alexander (*The Planet Uranus*) with additional material from Arago (*Popular Astronomy*) and Grant (*History of Physical Astronomy*).

4: AN ASTRONOMICAL MYSTERY

The various theories devised to explain the anomalous motion of Uranus are summarized by Nicol (*The Planet Neptune*) and Le Verrier (*Recherches*), who demolishes them one by one. The suggestion of "Ophion" is quoted by Nieto (*The Titius-Bode Law of Planetary Distances*). Valz's suggestion of a trans-Uranus planet is quoted by Grosser (*The Discovery of Neptune*). Airy's correspondence on the matter with Hussey and Eugene Bouvard can be found in his Neptune File. (Airy kept facsimile copies, made using some kind of bromide process, of all outgoing correspondence in this file; many of these copies have faded and are now almost illegible.) The character sketch of Airy is drawn from Maunder (*The Royal Observatory Greenwich*), Meadows (*The Royal Observatory*), and Wilfrid Airy's account of his father in *Autobiography*. Bessel's "vein of pure gold" speech is quoted from Herrmann (*History of Astronomy*), and Mädler's suggestion of a trans-Uranian planet is from Grosser.

5: THE YOUNG DETECTIVE

John Couch Adams's biographical details are drawn from Glaisher ("Biographical Notice"), Harrison (*Voyager*), Smart (*John Couch Adams*), and Jones (*John Couch Adams*). Details of Cambridge exam traditions are from Ball (*A History of the Study of Mathematics at Cambridge*). George Adams's "Reminiscences" are quoted from Harrison. Challis's letter to Airy requesting data on the position of Uranus on Adams's behalf is in the Neptune File. The description of Adams's mathematical approach is drawn from Grant (*History of Physical Astronomy*) and Adams ("Explanation" and *Scientific Papers*). Challis's letter of introduction to Airy, and Airy's reply, are in the Neptune File. Adams made his remark about "practical astronomers" in "Explanation." Details of Airy's daily routine are from Chapman (*Private Research and Public Duty*). Adams's folded-up note to Airy giving his prediction of the position of the new planet is in the Neptune File.

6: THE MASTER MATHEMATICIAN

Details of the Richardson case are from the *Times;* Airy's journal entry noting Richardson's suspension is quoted from Chapman (*Private Research and Public Duty*). Airy's reply to Adams is in the Neptune File. The question of the significance of the radius vector is discussed by Littlewood (*Mathematician's Miscellany*). Biographical details of Le Verrier are drawn from "Obituary" and *Centenaire de la Naissance,* Ball (*Great Astronomers*), Dunkin (*Obituary Notices of Astronomers*), Sheehan and Baum ("Vulcan Chasers" and *In Search of Planet Vulcan*), and the biography of Le Verrier on the Paris Observatory's website (http://www.obspm.fr/histoire/acteurs/leverrier.fr.shtml). The description of Le Verrier's mathematical approach is drawn from Grant (*History of Physical Astronomy*) and Le Verrier (*Recherches*). Airy's reaction to Le Verrier's second paper on Uranus is quoted from "Account," and his letter to Whewell is quoted from Smith (Cambridge Network). Airy's letter to Le Verrier about the radius vector, and Le Verrier's reply, are in the Neptune File. Airy's remarks to the Board of Visitors are quoted from "Account."

7: THE NOBLEST TRIUMPH OF THEORY

The discussion of Airy's motives for asking Challis to search for the new planet from Cambridge is drawn from Smith (Cambridge Network) and

Chapman (*Private Research and Public Duty*), but more closely follows Smith's interpretation. Airy's letters to Challis asking him to begin the search and suggesting an observational method are in the Neptune File. The description of the Cambridge search is from Challis ("Account"), from his correspondence with Airy in the Neptune File, and from his later letters to the *Cambridge Chronicle* and *The Athenaeum*, which appear as clippings in the Neptune File. Details of the efforts made in London by John Hind and at the Paris Observatory to search for the new planet are drawn from Smith; Maury's failure to launch a search in Washington, D.C., is quoted from Grosser (*The Discovery of Neptune*). Le Verrier's suggestion that the new planet could be identified by its disk is from *Recherches*. Adams's letter giving his second prediction of the planet's prediction is in the Neptune File. Schumacher's letter to Le Verrier is quoted by Gloden (*Centenaire de la découverte de Neptune*). Le Verrier's letter to Galle and the subsequent discovery of the planet by Galle and d'Arrest are described by Galle ("Ueber die Ertse Auffindung des Planeten Neptun"), Dreyer (*Historical Note*), Turner ("Obituary Notice of Johann Gottfried Galle"), and Wattenberg (*Johann Gottfried Galle*). Encke's quip about the visibility of Neptune's disk is quoted from a letter to the *Sidereal Messenger*. The letters from Encke and Schumacher congratulating Le Verrier are quoted from Grosser.

8: Possession of a New World

Galle's correspondence with Le Verrier is quoted in *Recherches*. Le Verrier's letter to Airy announcing the discovery of the new planet and naming it Neptune is in the Neptune File. The published statements made by John Herschel, Hind, and Challis are quoted from clippings in the Neptune File. Airy's correspondence with Challis and Le Verrier is in the Neptune File. Hind's complaint of a Cambridge conspiracy, White's "Cambridge snuggery" remark, and Brewster's similar suggestions are quoted from Smith (Cambridge Network). The description of the Royal Astronomical Society meeting is based on the accounts of Airy, Challis, and Adams that appeared in the Monthly Notices of the RAS. Airy's correspondence with Sedgwick and Le Verrier is in the Neptune File, as is the conspiratorial *Mechanics' Magazine* article.

9: An Elegant Resolution

John Herschel's sleepless night is quoted from Smith (Cambridge Network). His letter to the *Guardian* is quoted from clippings in the Neptune File, and his insistence that he was glad not to have discovered Neptune accidentally in 1830 is from a letter to Wilhelm Struve quoted by Buttman (*Shadow of the Telescope*). Smyth's remark to Airy about the naming of planets is quoted from Jones (*John Couch Adams*). Herschel's letter to Sheepshanks in which he says Neptune "ought to have been born an Englishman" is quoted from Smith. Herschel's "burn this" remark to Sheepshanks is quoted from a letter in the Herschel papers at the Royal Society. Airy's fence-mending correspondence with Challis and Adams is in the Neptune File. Hansen's belief that Adams's work was more mathematically beautiful than Le Verrier's is noted by Pannekoek (*The Discovery of Neptune*). Biot's "talented young man" remark is quoted from the *Journal des Savants,* by way of Grosser. The Russian opposition to the name "Le Verrier" for Neptune, on the grounds that it would be unfair to Adams, is noted by Challis ("Determination of the orbit of the planet Neptune"). Schumacher's use of the word "intolerable" is in a letter to Airy in the Neptune File. Adams's decision to turn down a knighthood is recounted by Harrison (*Voyager*). The controversy over Benjamin Pierce's claim that the discovery of Neptune was a "happy accident" is examined by Hubbell and Smith ("Neptune in America") and Gould (*Report on the History of the Discovery of Neptune*). The description of the meetings between Adams and Le Verrier is from Smart ("John Couch Adams and the Discovery of Neptune").

10: In Neptune's Sway

The description of the later life of Le Verrier is drawn from the same biographical sources as in chapter 6, with additional material about the "painful scene" from Dunkin (*Obituary Notices of Astronomers*) and Le Verrier's alleged reluctance to observe Neptune from Flammarion (*Popular Astronomy*). The account of the search for Vulcan is drawn from Hanson ("Leverrier, the Zenith and Nadir of Newtonian Mechanics") and Sheehan and Baum ("Vulcan Chasers" and *In Search of Planet Vulcan*). Details of Adams's later life, including his failure to prevent cheating in Cambridge exams and his letter to Sedgwick about marriage, are from Harrison (*Voy-*

ager) and Littman (*Planets Beyond*). The account of Airy's later life and flawed lunar calculations is from Airy (*Autobiography*) and Meadows (*The Royal Observatory*).

11: Shots in the Dark

Le Verrier's optimistic suggestion that further planets would be discovered using the same method as that used to find Neptune is quoted from *Recherches*. The various attempts to predict the positions of trans-Neptune planets are detailed by Gore (*Astronomical Curiosities*); the account of the search for Pluto is from Tombaugh and Moore (*Out of the Darkness*). The discussion of Pluto's dwindling mass and dubious planethood and the properties of trans-Neptunian objects is based on an interview with Brian Marsden of the Harvard-Smithsonian Center for Astrophysics. The early history of the search for extrasolar planets is drawn from Mammana and McCarthy (*Other Suns, Other Worlds?*) and Croswell (*Planet Quest*). The description of the search for planets around Barnard's Star is from van de Kamp ("Barnard's Star 1916–1976" and "The Planetary System of Barnard's Star"). One factor that helped to reignite interest in Planet X in the 1980s was the discovery that Galileo Galilei had seen Neptune and taken it for a star in December 1612 and January 1613, while observing the planet Jupiter through his telescope. (Galileo was thus the first person to see Neptune, though he did not realize what it was, while Adams, who was the first to detect Neptune, did so without seeing it.) Standish's determination of the nonexistence of Planet X based on Voyager 2's improved mass of Neptune was recounted by Marsden. At a party after the Voyager 2 flyby, Jurrie Van Der Woude, a member of the NASA mission team, was asked by Oliver Morton, a British science journalist, which of the four planets (Jupiter, Saturn, Uranus, and Neptune) visited by the probe he considered the most beautiful. Was it Jupiter, perhaps, with its giant moons and great red spot? "She is a painted harlot," came the reply. "Neptune is an Audrey Hepburn planet."

12: Worlds Unseen

The discussion of the modern discoveries of extrasolar planets, including the Upsilon Andromedae system, transits, naming, and distinguishing between small stars (brown dwarfs) and big planets is based on interviews and correspondence with Didier Queloz, Geoffrey Marcy, Greg Henry, Robert Noyes, Michael Nieto, and Brian Marsden, with additional material from Mammana and McCarthy (*Other Suns, Other Worlds?*) and Croswell (*Planet Quest*). The discussion of the future of planet hunting and the use of ground- and space-based interferometry is based on interviews with Didier Queloz, Michael Shao, and Alan Penny. Some readers may wonder why I have chosen to ignore the discovery of planetary-mass companions orbiting a pulsar in 1991. My reasoning is that such bodies (and others discovered subsequently) are not true planets because they are most likely to have formed via some unknown and exotic process, rather than from an accretion disk around a young star (which I take to be the traditional definition of a planet). This suggests that pulsar planets can tell us very little about planetary formation and planetary systems in general, and makes 51 Pegasi's planet the first "true" extrasolar planet to have been discovered.

Sources

Manuscript Collections

Airy, George. "Neptune File," Royal Greenwich Observatory archive, University Library, Cambridge.

Herschel, John. Papers. Royal Astronomical Society, London.

Books and Articles

Adams, John Couch. "An Explanation of the Observed Irregularities in the Motion of Uranus, on the Hypothesis of a Disturbance Caused by a More Distant Planet; with a Determination of the Mass, Orbit and Position of the Disturbing Body." *Monthly Notices of the Royal Astronomical Society* 7 (1846): 149–52.

Adams, W. G., ed. *The Scientific Papers of John Couch Adams.* Cambridge: Cambridge University Press, 1896.

Airy, George Biddell. "Account of Some Circumstances Historically Connected with the Discovery of the Planet Exterior to Uranus." *Monthly Notices of the Royal Astronomical Society* 7 (1846): 121–44.

Airy, Wilfrid, ed. *Autobiography of Sir George Biddell Airy.* Cambridge, 1896.

Alexander, Arthur Francis O'Donel. *The Planet Uranus: A History of Observation, Theory and Discovery.* London: Faber and Faber, 1965.

Armitage, Angus. *William Herschel.* London: Thomas Nelson and Sons, 1962.

Arago, François. *Popular Astronomy.* Trans. W. H. Smyth and R. Grant. London: Longman, Brown, Green and Longmans, 1855.

———. *Popular Lectures on Astronomy.* New York: Greeley and McElrath, 1848.

Ashworth, William. "Herschel, Airy and the State." *History of Science* 36, no. 112 (1998): 151–78.

Ball, Robert. *Great Astronomers.* London: Sir Isaac Pitman and Sons, 1895.

Ball, Walter William Rouse. *A History of the Study of Mathematics at Cambridge.* Cambridge: Cambridge University Press, 1889.

Baum, Richard, and Sheehan, William. *In Search of Planet Vulcan: The Ghost in Newton's Clockwork Universe.* New York: Plenum, 1997.

Bell, E. T. *Men of Mathematics.* New York: Simon and Schuster, 1937.

Buttmann, Gunther. *The Shadow of the Telescope: A Biography of John Herschel.* New York: Charles Scribner's Sons, 1970.

Challis, James. "Account of Observations at the Cambridge Observatory for Detecting the Planet Exterior to Uranus." *Monthly Notices of the Royal Astronomical Society* 7 (1846): 145–49.

———. "Determination of the Orbit of the Planet Neptune." *Astronomische Nachrichten* 596 (1847): 309–14.

Chapman, Allan. *Dividing the Circle: The Development of Critical Angular Measurement in Astronomy, 1500–1850.* New York: E. Horwood, 1990.

———. "Private Research and Public Duty: George Biddell Airy and the Search for Neptune." *Journal for the History of Astronomy* 19, (1988): 121–39.

Clerke, Agnes M. *The Herschels and Modern Astronomy.* London: Cassell, 1895.

———. *A Popular History of Astronomy During the Nineteenth Century.* New York: Macmillan, 1886.

Croswell, Ken. *Planet Quest: The Epic Discovery of Alien Solar Systems.* San Diego: Harcourt Brace, 1997.

Cunningham, Clifford J. "The Baron and His Celestial Police." *Sky and Telescope*, March 1988.

———. "Giuseppe Piazzi and the Missing Planet." *Sky and Telescope,* September 1992.

Dreyer, J. L. E. "Historical Note Concerning the Discovery of Neptune." *Copernicus* 2 (1882): 63–64.

Dreyer, J. I E., and Turner, H. H., eds. *History of the Royal Astronomical Society, 1820–1920.* London: Royal Astronomical Society, 1923.

Dunkin, Edwin. *Obituary Notices of Astronomers.* London: Williams and Norgate, 1879.

Dunnington, G. Waldo. *Carl Friedrich Gauss: Titan of Science.* New York: Exposition Press, 1955.

Encyclopaedia Britannica. 11th ed. Cambridge: Cambrige University Press, 1911.

Flammarion, Camille. *Popular Astronomy.* 2nd ed. London: Chatto and Windus, 1907.

Galle, Johann Gottfried. "Ueber die Erste Auffindung des Planeten Neptun." *Copernicus* 2 (1882): 96–97.

Glaisher, J. W. L. "Biographical Notice of John Couch Adams." In *The Scientific Papers of John Couch Adams*. Cambridge: Cambridge University Press, 1896.

Gloden, Albert. *Le Centenaire de la découverte de Neptune par Le Verrier.* Paris: Gauthier-Villars, 1947.

Gore, J. Ellard. *Astronomical Curiosities.* London: Chatto and Windus, 1909.

Gould, Benjamin Apthorp. *Report on the History of the Discovery of Neptune.* Washington, D.C.: Smithsonian Institution, 1850.

Grant, Robert. *History of Physical Astronomy.* London: Henry G. Bohn, 1852.

Grosser, Morton. *The Discovery of Neptune.* Cambridge, Mass.: Harvard University Press, 1962.

Hanson, N. "Leverrier, the Zenith and Nadir of Newtonian Mechanics." *Isis* 53 (1962): 359.

Harrison, H. M. *Voyager in Time and Space: The Life of John Couch Adams, Cambridge Astronomer.* Sussex: Book Guild, 1994.

Herrmann, Dieter B. *The History of Astronomy from Herschel to Hertzsprung.* Cambridge: Cambridge University Press, 1984.

Herschel, John. *Outlines of Astronomy.* London: Longman, Brown, Green and Longmans, 1850.

Herschel, William. "Account of a Comet." *Philosophical Transactions of the Royal Society* 71 (1781): 492–501.

Holden, Edward Singleton. *Sir William Herschel : His Life and Works.* London: W. H. Allen, 1881.

Hubbell, John G., and Smith, Robert W. "Neptune in America." *Journal for the History of Astronomy* 23 (1992): 261–291.

Hyman, Anthony. *Charles Babbage, Pioneer of the Computer.* Oxford: Oxford University Press, 1982.

Institut de France. *Centenaire de la naissance de Urbain Jean-Joseph Le Verrier.* Paris: Gauthier-Villars, 1911.

Jackson, J. "The Discovery of Neptune: A Defence of Challis." *Monthly Notices of the Royal Astronomical Society of South Africa* 8 (1949): 88–89.

Jones, Harold Spencer. *John Couch Adams and the Discovery of Neptune.* Cambridge: Cambridge University Press, 1947.

Kowal, Charles T., and Drake, Stillman. "Galileo's Observations of Neptune." *Scientific American,* December 1980 pp. 52–59.

Langley, Samuel Pierpont. *The New Astronomy.* Boston: Ticknor, 1888.

Le Verrier, Urbain Jean-Joseph. "Recherches sur les Mouvements de la Planète Herschel, dite Uranus" (Paris, 1846, pp. 3–254), published as an extract in *Connaissance des temps pour 1849.*

Littlewood, J. E. *A Mathematician's Miscellany.* London: Methuen, 1953.

Littman, Mark. *Planets Beyond.* New York: Wiley, 1988.

Lubbock, Constance A., ed. *The Herschel Chronicle: The Life-Story of William Herschel and His Sister, Caroline Herschel.* New York: Macmillan, 1933.

Macpherson, Hector. *Makers of Astronomy.* Oxford: Clarendon Press, 1933.

Mammana, D. L., and McCarthy, D. W., Jr. *Other Suns, Other Worlds?* New York: St Martin's Press, 1996.

Maunder, E. Walter. *The Royal Observatory Greenwich.* London: Religious Tract Society, 1900.

McCrea, William Hunter. *The Royal Greenwich Observatory: An Historical Review Issued on the Occasion of Its Tercentenary.* London: HMSO, 1975.

Meadows, A. J. *The Royal Observatory at Greenwich and Herstmonceux, 1675–1975.* Vol. 2, *Recent History (1836–1975).* London: Taylor and Francis, 1975.

Mitton, Jacqueline. *Dictionary of Astronomy.* London: Penguin Books, 1998.

Moore, Patrick. *The Planet Neptune: An Historical Survey Before Voyager.* Chichester: J. Wiley, 1996.

———. *William Herschel: Astronomer and Musician of 19 New King Street, Bath.* Sidcup: P. M. E. Erwood / the William Herschel Society, 1981.

Nicol, J. P. *The Planet Neptune, an Exposition and History.* Edinburgh: J. Johnstone, 1848.

Nieto, Michael Martin. *The Titius-Bode Law of Planetary Distances.* New York: Pergamon, 1972.

"Obituary of Urbain Le Verrier." In *Report of the Council to the 58th Annual General Meeting. Monthly Notices of the Royal Astronomical Society* 38 (1878): 155–66.

Pannekoek, A. "The Discovery of Neptune." *Centaurus* 3 (1953): 126–37.

Planche, J. R. *The New Planet, or Harlequin out of Place, an Extravaganza in One Act.* London: S. G. Fairbrother, 1847.

Porter, Roy, ed. *Hutchinson Dictionary of Scientific Biography.* Oxford: Helicon, 1994.

Schaffer, Simon. "Uranus and Herschel's Astronomy." *Journal for the History of Astronomy* 12 (1981).

Sheehan, William, and Baum, Richard. "Vulcan Chasers." *Astronomy,* December 1997.

Sidgwick, J. B. *William Herschel, Explorer of the Heavens.* London: Faber and Faber, 1953.

Smart, W. M. "John Couch Adams and the Discovery of Neptune." *Occasional Notes of the Royal Astronomical Society* 11, no. 2, (1947): 33–88.

Smith, Robert W. "The Cambridge Network in Action: The Discovery of Neptune." *Isis* 80 (1989): 395–422.

Teets, Donald, and Whitehead, Karen. "The Discovery of Ceres: How Gauss Became Famous." *Mathematics Magazine,* April 1999 pp. 83ff.

"The Telescopic Discovery of Neptune." *Journal of the British Astronomical Association* 61 (1951): 166.

Tombaugh, Clyde W, and Moore, Patrick. *Out of the Darkness: The Planet Pluto.* Harrisburg, Penn.: Stackpole Books, 1980.

Turner, Herbert Hall. "Obituary Notice of Johann Gottfried Galle." *Monthly Notices of the Royal Astronomical Society* 71 (1911): 275–81.

van de Kamp, Peter. "Barnard's Star, 1916–1976: A Sexagintennial Report." *Vistas in Astronomy* 20 (1977): 501.

———. "The Planetary System of Barnard's Star." *Vistas in Astronomy* 26 (1983): 141.

von Zach, Franz Xaver. "Ueber den neuen Hauptplaneten." *Montaliche Correspondenz zur Beforderung der Erd und Himmels-Kunde* 3 (1801): 592–623.

Wattenberg, Diedrich. *Johann Gottfried Galle: Leben und Wirken eines Deutschen Astronomen.* Leipzig: Johann Ambrosius Barth, 1963.

PERIODICALS

Athenaeum (London): October 3, 15, 17, November 21, 1846; February 20, 1847.

Illustrated London News (London): October 10, 24, 1846; January 2, 1847.

Scientific American (New York): November 21, December 19, 1846; February 20, March 20, 27, May 22, June 12, 1847.

Sidereal Messenger (Cincinnati): December 1846; January 1847.

Times (London): January 26, 31, February 3, 6, 7, 13, 16, 24, May 14, October 1, 1846; March 15, 1847.

Index